THE CLASSIC WATCH

THE CLASSIC WATCH

MICHAEL BALFOUR

Quantum Books

A QUANTUM BOOK

This book is produced by
Quantum Publishing Ltd.
6 Blundell Street
London N7 9BH

This edition printed 2005

ISBN 1-84573-096-8

QUMWAT

Printed in Singapore by
Star Standard Industries Pte Ltd

C O N T E N T S

TO CALANDRA
who kept perfect time on
17 December, 1978

INTRODUCTION

MOST OF US WEAR ONE. Most of us believe we couldn't do without one, that we would be 'lost' without a wristwatch to confirm the second in the minute in the hour in every day. It is said that as many people wear wristwatches as there are people wearing shoes on their feet, and so the market for them is massive and ever-expanding.

The majority of the market is satisfied with an inexpensive, uncomplicated quartz digital or analog watch on its wrist, just as long as it 'keeps time'. In recent years, however, the most interesting and unusual wristwatches have become recognized as collectable, and now investors and collectors alike are appreciating the financial gains as well as the pleasures of these miniature, yet intricate, machines. Just as Rolls-Royce, Bentley, Ferrari, Bugatti and other famous motorcar marques feature most prominently in auctions, showrooms and dealers' lists, so also a small number of watchmakers attract the most attention from dealers and collectors. This book is an introduction to the manufacturers of past, present and future classic wristwatches, and their greatest achievements.

HOW TO USE THIS BOOK

The watch trade has its own language, and it is essential to understand the meanings of key words; in the glossary at the end of the book under the letter 'b', for example, are grouped such words as back winder, baguette, balance, band, barrel, baton, bezel, bi-metallic and bridge.

The glossary gives the precise meanings of over 100 words and phrases which a good auction catalog will use. These words are a form of shorthand with which the serious reader, who intends to buy, invest or collect, should become familiar.

There are two main parts to the contents of this book. The first provides readers with an 'A to Z' of the leading manufacturers of wristwatches and their most collectable models. These watches are generally regarded as classic, of enduring appeal, for one or more of very few reasons which are discussed: they involve beauty, design, technical innovation and complication.

The other major part of the book deals with collecting. It is important to grasp the basic principles by which to be guided, whether you are buying privately, from dealers, from jewelers or through auction salerooms and their catalogs. Having acquired a wristwatch you will need to know how to keep it, look after it, store it, and indeed insure it. Swiss watch trade organizations and the larger individual manufacturers devote a large proportion of their annual expenditure budgets to tracking down and shutting down makers and pedlars of fake wristwatches: these necessary policies, however, can unhappily never completely succeed. Citizen have recently begun incorporating a hologram seal in their watches to help identify fakes.

Most collectors specialize: a world stamp collection is surely one upon which only the very rich or the very foolish (or an unsubtle combination of both characteristics) would embark. Thus also with wristwatches, and so here you will find suggestions for some basic themes for collections; some of them will be expensive, but certainly not all, and anyway you probably will not be contemplating a collection unless you know you are going to find real bargains! After all, we all know somebody who has, don't we? When you have chosen a theme for your future collection, you will have direction and you will soon know more or less exactly which wristwatch models to look out for: the hard way to do this is to sit at home and wait for telephone calls from the few people you have told about them. The easier way is to use the list of Useful Addresses at the back of this book, and contact all of them.

OPPOSITE, ABOVE AND BELOW *Patek Philippe's famous Calibre 89, which fetched 4.9m Swiss francs at a specialist auction of their watches held by Habsburg, Feldman, Geneva, on April 9, 1989. At that time it became the most expensive watch in the world; it was acquired by an investment consortium who promptly named it the Kuma, after a Japanese goddess.*

OPPOSITE, CENTER *A selection of watches auctioned at Christie's, London, on January 22, 1988. The Rolex Oyster Cosmograph (top row, 3rd from right) fetched one of the highest prices of the day.*

ABOVE *This Russian world timer, called Paketa, is a cult watch for young collectors in the West.*

EARLY TIME RECORDERS

It is beyond the scope of this book to explore the very nature of time, or to describe in any detail the origins and development of devices for the recording and displaying of the passing of time. A subject must be vast which is involved with the orbital motion of Earth; it weighs six thousand six hundred million million million tons, it spins on a 7,926-mile axis, and travels through space at 18,520 miles per hour.

Prehistoric man started to build Stonehenge around 3000 BC; the task was left uncompleted around 1100 BC. Through its six different stages of construction, the intentions of its builders changed, but at all times it was a calendrical instrument of most remarkable precision. It was also an eclipse predicter – important then, but of no concern to wristwatch designers today. During the last millennium BC, water clocks were evolved by Chinese, Egyptians and Greeks, whilst the Japanese had their incense clocks. The Mayan calendar was evolved between AD 300 and 900. In Britain King Alfred (849–99) was famous for his candle clock. The Islamic culture developed the sciences of mathematics and astronomy, and in *c.* AD 950 the astrolabe was evolved. Sun-dials for outside and hour glasses for inside the home were common. In the 15th century the Incas paid great attention to time and its divisions, while at the same time, in Europe, cathedrals and churches were acquiring their wall-mounted dials and early clocks. Copernicus (1473–1543) lent his heliocentric theory to the development of the clock, and Galileo (1564–1642) proved that the length of a pendulum, not its amplitude, dictates the time of its swing; the first mention of a mechanical clock in Japan occurs soon after 1550. Around 1500 Peter Henlein of Nuremberg is credited with inventing the mainspring, and clocks became portable. Thomas Tompion (1639–1713) was one of the greatest of all English clockmakers, and he helped to develop lantern, longcase, bracket and carriage clocks.

Between 1770 and 1840 France dominated clock and pocket watchmaking; the greatest maker at this time was Abraham-Louis Breguet (1747–1823), who invented the tourbillon and also the Tact watch for the blind. In 1770 Jean-Antoine Lepine devised a new watch movement so that watches could be made much thinner; he also introduced hand setting at the back of the watch, a case with invisible hinges (when closed) and, within the movement, wheels with 'wolf's teeth'. At this time watchmaking countries such as Austria, Italy, Scandinavia, Spain and Switzerland were being influenced in all they did by France.

BOTTOM *This 1892 advertisement is one of the earliest to depict a wristwatch; it is described as a 'fine gold keyless watch bracelet'.*

RIGHT: *The ancient clock in Salisbury Cathedral.*

'SWISS MADE'

These familiar words are to be found on the dials of wristwatches made in Switzerland – or, in these days of fakes, supposedly Swiss-made, whatever their quality. In 1601 the Watchmakers' Guild of Geneva was formed, and the industry began to grow; within a century there were 500 watchmakers in the city. In 1685 came the revocation in France of the Edict of Nantes, which had guaranteed the freedom of Protestants in 1598; the harsh political and religious effects of this drove hundreds of Huguenot watchmakers south into the Jura Mountains in and around La Chaux-de-Fonds in northern Switzerland. In about 1790 Jaquet Droz of Neuchâtel is said to have designed the first ever wristwatch. At this time the numbers of 'cabinotiers' were increasing fast in the Jura villages, making separate bits and parts and assembling pocket watches and clocks by hand. In 1808, 12 per cent of Geneva's population of 30,000 were involved in watchmaking; in the canton of Neuchâtel no less than 400 watchmakers were registered – a fact that demonstrates how the Swiss watchmaking industry has always been conducted in a highly organized fashion.

Now came some key developments that were to lead directly to the wristwatch. In 1837 the Geneva craftsman Antoine Leschaud refined a recent English invention into the jeweled lever escapement. A year later, Louis Audemars invented the stem winding and setting mechanism. In 1868, the year in which Patek Philippe first made a wristwatch, Georges-Frédéric Roskopf developed his pin-lever movement. In 1871 Aaron Dennison (see International Watch Company in the 'A to Z of Important Makers') in America invented the waterproof watch case with screw button and pendant, while Georges Leschot was devising his system of interchangeable parts which led to series production. In 1880 Girard-Perregaux produced its famous watch for German naval officers, and its familiar four by four metal protective grille stares out of the record books as the world's first true production watch. It has often been said that Omegas were used in the Second Boer War (1889–1902), but this seems not to have been proved as a fact.

From this point on the story of 'Swiss Made' watches is told through the profiles of the important makers which follow this Introduction. The stage is now set for the wristwatch industry to come into being.

The bottom center watch in the picture opposite, shown in greater detail. It is a Vacheron & Constantin watch made for Bulgari c. 1950. Curved asymmetrical cases are popular among collectors.

OPPOSITE, ABOVE *Watches auctioned at Sotheby's, London, December 13, 1988. Second from left, top row, is a Cartier 18 karat gold and enamel lady's wristwatch made in about 1920. The signed movement has 18 jewels, the engine-turned dial has Arabic numerals, and the bezel features two bands of white enamel.*

OPPOSITE, BELOW *More watches from the same sale: among the popular watches depicted are the 1936 18 karat Movado (top left), the 1929 18 karat Rolex Prince (top center). One of the highest prices of the day was obtained for the three-color gold Rolex Datejust Oyster Perpetual, with its distinctive champagne dial, diamond numerals and clear calendar aperture.*

THE LANGUAGE OF A WRISTWATCH

In this book several different kinds of wristwatches are discussed; for a fuller description of these, reference should be made to the Glossary of Terms.

A mechanical watch (which by definition must have more than 120 parts) is powered by a mainspring within what is called a barrel: the mainspring is wound by hand or automatically by a rotor, transmitting power to a complicated set of interconnected wheels; this is called the gear train. The train in turn is connected to the escapement mechanism, which consists of a wheel and pallet fork, which, again in turn, transmits impulses to the balance wheel, making it oscillate. This oscillation provides the 'tick' of the mechanical watch. There is a further train of interconnected wheels carrying the hands indicating the time. The rotor on an automatic or self-winding mechanism incorporates a weight, which swings backwards and forwards whenever the wrist moves. These oscillations are converted into a rotary motion with a gear train which winds the mainspring. Over 1,850 different operations are involved in making a mechanical wristwatch and over 1,000 tools are used.

There are various contemporary variants on the traditional hand-wound mechanism. Automatic watches, for instance, rely on wrist and arm movement for winding power, though this assumes they are worn constantly. They are also designed in such a way that it is impossible to over-wind them. An electronic wristwatch is powered by a quartz crystal. Nowadays these crystals are grown under laboratory conditions to ensure greater purity. The principal components of a quartz watch are a battery, a step-motor, a quartz resonator and an integrated circuit. A quartz watch may be either an analog watch, which has a traditional dial and hands, or a digital watch, which displays the time in numbers in liquid crystal. When activated by the power of the battery, the incorporated quartz vibrates extraordinarily rapidly, as it divides the passing of time into 32,768 equal parts per second. This very high frequency, as it is called, is the reason why a quartz watch will always be about 12 times more accurate than a simpler mechanical version. Quartz crystals scarcely age because they are specially treated during their artificial manufacture; moreover they are not affected by changes in atmospheric pressure or extreme shocks. A major source of natural quartz crystals is Brazil.

Descriptions of case shapes, measurements, dials, apertures, numerals, hands, and so on, will be found in the Glossary of Terms. The vast majority of wristwatches sold around the world are made of stainless steel. The most expensive watches are most often made of 14, 18 or 22 carat yellow, pink or white gold, though the wristwatch buyers' and collectors' market of today is naturally affected by the Muslim decree that precious metals should not be worn. They may also be made of platinum, vermeil (silver gilt), or titanium. In late 1988 Audemars Piguet introduced the first tantalum watch (automatic, three subsidiary dials, calendar, chronograph); tantalum is blue-gray in color (in the mass, so scratches do not matter); it is almost as hard as steel (although 200 times more expensive) while its density is only 10 per cent less than pure gold. The rarity of the basic raw material means that very few of these watches will be made; they will undoubtedly become collectors' pieces.

Elsewhere, it is explained why it is not wise to open up a wristwatch unless absolutely necessary. The best watches are made to be utterly secure from any intrusion. It is interesting here to note that the word 'waterproof' is not used; the correct expression is 'water-resistant'. Water-resistance is measured in 'atmospheres', and most water-resistant wristwatches are tested to three atmospheres – that is, twice the normal atmospheric pressure. Unless specifically made for swimming and diving (Tag-Heuer and Citizen, for instance, have such special ranges), the average water-resistant watch is not intended to be immersed in water; hot water can distort the glass and washers. According to international standard specifications a watch may be called water-resistant if it survives immersion under a meter (3.28 ft) of water for 30 minutes and immediately afterwards under 20 meters (65.6 ft) for 90 seconds.

One easy way to learn the language of the wristwatch is by reading through auction catalogs; the accurate and precise descriptions will help make you familiar with the terminology.

ABOVE, TOP TO BOTTOM
A gold Universal Tri-Compax calendar, moonphase and chronograph; an unusual Chronoswiss, Kelek five-minute repeating automatic; an Omega Constellation automatic with calendar.
RIGHT, TOP TO BOTTOM
An 18 karat Rolex Oyster Perpetual Datejust, 1961; an 18 karat Vacheron & Constantin automatic calendar wristwatch; a stainless steel Omega Flightmaster chronograph; a stainless steel Omega Speedmaster Professional.

FAR RIGHT:
This recent Sectora range has been produced by Jean d'Eve and might become collectable in the future. The time can be read vertically over an angle of 150°.

WHAT IS A CLASSIC WRISTWATCH?

A wristwatch makes a statement about its wearer. It is not a necessary adornment, and yet it is commonly worn; it has been bought with great forethought and care, whether for the purchaser or for someone else, out of duty or love; an image of the wearer has been called up and, with luck, matched to the design, shape, color, functions and cost of an exquisite machine, which may then be worn and consulted daily for decades to come.

With the possible exception of cuff-links, signet and wedding rings, wristwatches are the only items of 'jewelry' which many men feel comfortable wearing. In general, it appears that wristwatch collecting is a male pursuit, which reveals a throwback to the centuries-old close association between jewelry, clocks and watches. It is commonly said that women do not like wearing an expensive secondhand watch, as they know it probably once belonged to another woman. Secondhand ladies' watches in auction salerooms sell mainly to trade dealers and collectors.

For ease of reference, classic wristwatches can be divided into three categories. The first is that of 'complicated' wristwatches. There are six classic kinds: the chronometer, the minute repeater (sometimes known as the bankers' watch), the moonphase calendar, the perpetual calendar, the tourbillon regulator and the ultra-thin chronograph (a stop-watch mechanism). Everybody has their own list of complicated classics; the following watches should find their way into such lists. From Patek Philippe, the 18 karat split-second chronograph, the 18 karat round moonphase chronograph, the World Time and the moonphase chronograph; the Vacheron & Constantin 18 karat minute repeater (*c.* 1935); International Watch Company's Da Vinci; the Astrolabium from Ulysse Nardin; the Audemars Piguet self-winding rectangular; and Longines' Lindberg Aviator.

Contemporary quartz Mickey Mouse watches from Apollo Watch Products, which reflect the continuing popularity of Walt Disney's most famous cartoon character.

The second category of classic wristwatches features those with unique designs. Informed lists might include the following: two watches from Jaeger LeCoultre, the Reverso and Diamond Mystery; from Rolex the flared case Prince (the doctor's watch) and the hooded two-tone bubbleback; the Movado Polyplan; from Hamilton the Flight I and Alistair; from Gruen, the drivers' watch and the Curvex General; Audemars Piguet's rectangular moonphase calendar skeleton; Salvador Dali's designs for Cartier and of course their Tank and Santos; the Illinois Piccadilly; Ingersoll's Mickey Mouse (1933; first watch for children).

A third category would be unusual and/or firsts. Ingersoll's Mickey Mouse could also be on this list, together with Waltham's masonic, Cartier's mammoth digital, Corum's Golden Bridge, Rolex's jump-hour Prince and the Bulova Accutron (1960; electronic tuning-fork watch invented by Max Hetzel); Harwood's automatic (1928); the Rolex water-resistant Oyster (1926) and automatic (1931); Breitling's first wristwatch chronograph (1914); Lip's 1953 battery-powered watch; Tissot's Rock Watch; Swatch's Jellyfish; and International Watch Company's Ingenieur.

In the 'A to Z of Important Makers', which follows this Introduction, the selection is necessarily the author's own; no official classification exists (as among French wine-growers, for instance). Several well-known makers have had to be excluded, for the lack of space, but many of their best-known models are mentioned in 'Collecting Wristwatches' later in the book.

RIGHT, ABOVE *This 1929 advertisement carries copy which is entirely justified; any wrist movement activated the winding mechanism. This remarkable early English invention by John Harwood in 1924 is perpetuated today in an elegant contemporary range of Harwood watches manufactured by Fortis.*

RIGHT, BELOW *Charles A Lindbergh (1902–74) made the first solo non-stop flight across the Atlantic Ocean from New York to Paris in 1927, in his monoplane Spirit of St Louis. He commissioned a watch which gave the time in degrees of the arc, thus making it easier to read the longitude. The Lindbergh Hour Angle watch has been available from Longines since 1932.*

OVERLEAF The Persistence of Memory *by Salvador Dali (1904–89): a surrealist vision of time.*

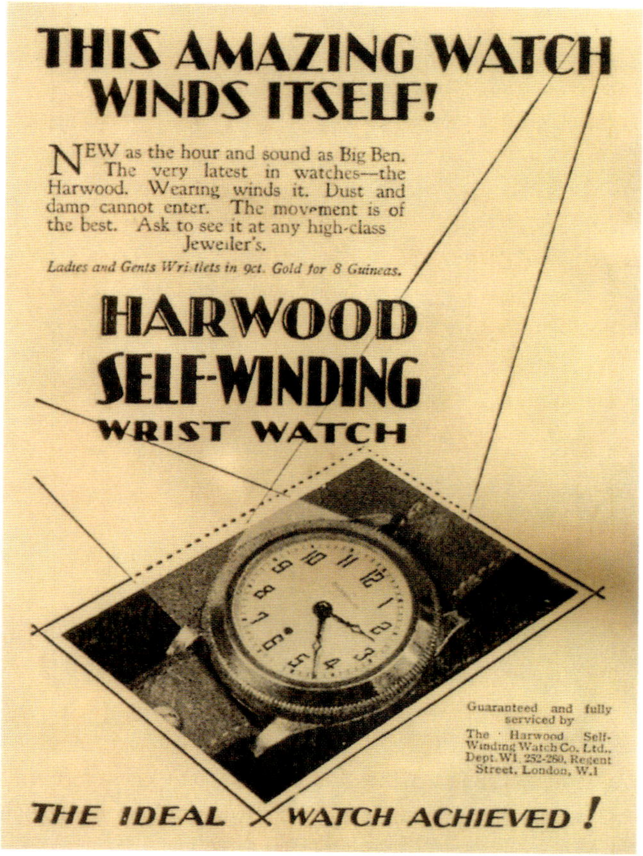
THIS AMAZING WATCH
WINDS ITSELF!
NEW as the hour and sound as Big Ben. The very latest in watches—the Harwood. Wearing winds it. Dust and damp cannot enter. The movement is of the best. Ask to see it at any high-class Jeweller's.
Ladies and Gents Wristlets in 9ct. Gold for 8 Guineas.
HARWOOD
SELF-WINDING
WRIST WATCH
Guaranteed and fully serviced by
The Harwood Self-Winding Watch Co. Ltd., Dept. W1. 252-260, Regent Street, London, W.1
THE IDEAL WATCH ACHIEVED!

TRIOMPHE
LINDBERGH

THE WRISTWATCH INDUSTRY TODAY

In 1988, 653 million wristwatches of all kinds were manufactured (according to a survey by Citizen). Of these 123 million were mechanicals, 228 million were digitals and 308 million were analog watches. The attraction of digital watches is apparently waning rapidly. Of the watches made, 24 per cent went to Western Europe, 22 per cent to North America, 10 per cent to China, 9 per cent to Eastern Europe, 6 per cent to Japan and 6 per cent to the Middle and Near East. The biggest maker of wristwatches was Japan (260 million), followed by Hong Kong (170 million; all digitals), and Switzerland (75 million). As in many industries, Japan's export achievements dominate. The Japan Clock and Watch Association estimate that the production costs of their analog watches are no less than 46 per cent lower than they were three years ago. The early days of attractive quartz digitals swamping the market are more or less over, and the attention in Japan is now turning to two other critical factors – complication and dial design. It is interesting to note that, conversely, Swiss watches increased their share of the Japanese market from 18 per cent to 20 per cent in 1988. As a measure of Japanese intentions towards the wristwatch market, Citizen have a research and development staff of about 400 people, to whom some 5 per cent of the value of their gross sales is allocated. In April 1989, at the Basle European Watch and Jewelry Fair, Citizen showed the world's first quartz analog wristwatch to incorporate a 200-year pre-set calendar; it has eight hands indicating the year, month, day, date, 24 hours, hour, minute and second.

LEFT *Probably the most successful attempt to capture the inexpensive quartz fashion wristwatch market has been achieved by Swatch, part of the giant SMH group. These are three of their range of many hundreds of stylish pop watches.*

RIGHT: *A precursor of the Swatch success was the French company Lip: this is Roger Tallon's famous Mach 2000 design.*

Brand marketing, again, is as important as it was 50 years ago, the difference being that now very few of the 'names' manufacture every part of a wristwatch. There are 'assemblers' at the bottom end of the market; the 'own brand' business continues to expand rapidly; Endura is one of the leaders in this field. When the invasion of quartz-powered movements overtook Switzerland at the beginning of the 1970s, many manufacturers disposed of their machinery for making mechanical watches. Oris, among others, were lucky or fortunate enough to retain their plant, and are thus in a position economically to re-enter the market for mechanicals – a market which has never really vanished. At the time of the quartz invasion many great watch names disappeared; many more were forced to sell out to larger groups; today in Switzerland, SMH owns Certina, Endura, Hamilton, Longines, Mido, Rado, Swatch and Tissot. In Japan the family-controlled Hattori own Jean Lassale, Lorus, Pulsar, Seiko and Yema.

The words 'Swiss Made' remain a guarantee today of a fine wristwatch. The watchmaking centers remain, as they always have, in Swiss cities, towns and villages like Bienne, Geneva, Grenchen, La-Chaux-de-Fonds, Le Brassus, Le Locle, Le Sentier, Lausanne, Neuchâtel, Schaffhausen, and Zurich. There is no longer a watchmaking industry of any size in America or the UK; Pforzheim in West Germany remains a traditional manufacturing center.

THE FUTURE OF WRISTWATCHES

This is assured, as annual wristwatch sales advance towards an expected 800 million per year in 1990. It is reasonable to predict, however, that cultivated attitudes toward wristwatches will alter, to the extent that each of us will own several watches, for different occasions (business, evening dress, sports, casual weekend). The market for children's watches will grow, and expensive models for the pampered child will surely arrive soon. Already, low-priced novelty watches either manifest, or produce new collecting habits or cults. The market for expensive mechanical analogs will never cease to expand. Great car and fashion designers (Enzo Ferrari, Yves Saint Laurent) are justly famous; great watch designers, such as Gērald Genta, will become more widely recognized for their art. It is a firm prediction that sooner or later an international de-luxe fashion goods and accessories firm will produce a classic watch; Gucci (and Hermès, with its 200 shops worldwide) are intimately involved with the manufacture of their wristwatches.

THE A-Z OF MAJOR MAKERS

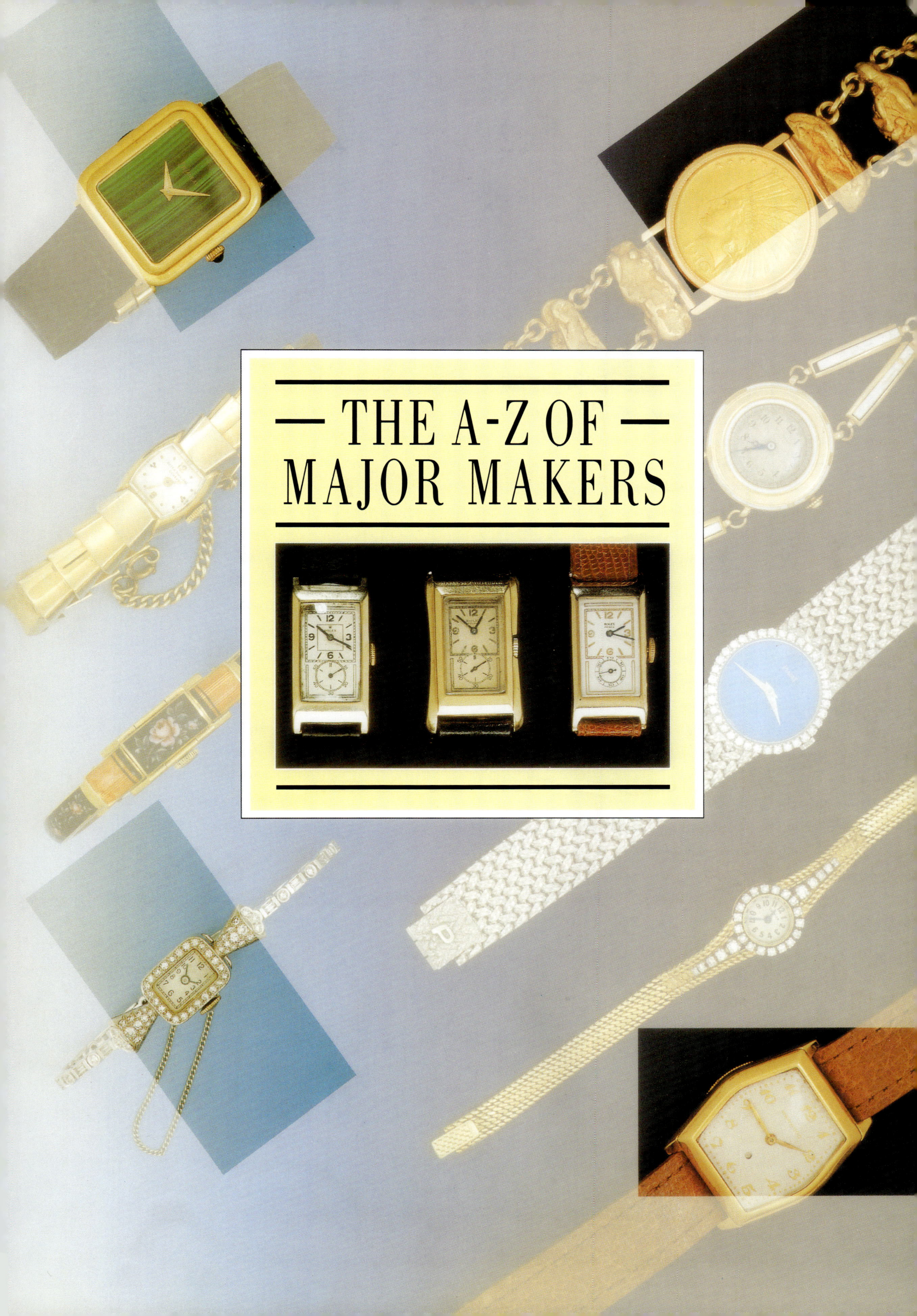

AUDEMARS PIGUET
60 TACHYMETER
AUDEMARS PIGUET
AUTOMATIC

AUDEMARS PIGUET

THIS DUAL NAME is on the dials of some of the greatest classic watches ever to be designed and constructed, and, for many enthusiasts, it has the same resonance as the words Rolls-Royce do for automotive buffs. Indeed there are many fascinating parallels to be found within the contexts of the two meetings between Jules Audemars (1851–1918) and Edouard Piguet (1853–1919) in Le Brassus in the Swiss Jura mountains in 1875, and between Sir Charles Rolls and Henry Royce in England at the turn of the century. It is never a surprise to find an Audemars Piguet on the wrist of a Rolls-Royce owner; such a purchaser undoubtedly appreciates beauty in movement.

At the time of their meeting in 1875, Audemars was a maker of dial frames and Piguet was just starting as a finisher. Apparently they started collaborating immediately. Their trademark was officially registered in December 1882, and by 1889 Audemars Piguet & Cie SA was in business, with ten employees, making both simple and complicated unsigned movements for other local watchmakers and for export to America, where the import tax on complete watches was then very high. The two gifted pioneers continued their business together until 1918, when Jules Audemars died; his partner died the following year. To this day the company has remained in the hands of the descendants of the founders, together with those of a few other original investors in La Vallée de Joux.

LEFT *A fine Audemars Piguet Automatic chronograph in 18 karat pink gold and tantalum.*

ABOVE *This Audemars Piguet has a distinctive tonneau-shaped textured case; the brushed gilt dial bears raised baton numerals and gold hands. The gold bracelet is integral with the case.*

Audemars Piguet has always maintained a detailed register of every watch made and sold; thus, as with Rolls-Royce chassis numbers, each product has a unique serial number, which makes a fake watch difficult to pass off (unless to unwary or uncaring buyers in, say, Hong Kong, to whom a ridiculously huge saving is all).

In its early years, when wristwatches were in the first stages of development, the firm naturally concentrated on pocket watches. Today some of them command astonishingly high prices in the auction rooms. The most famous of these are La Grande Complication, of which only 100 have been made between 1915 and 1989; the Grande Sonnerie, the Perpetual Calendar Watch with Minute Repeater, and the Skeleton Watch. However, after the First World War had finally established the wristwatch industry, most Swiss makers turned their attention to the worldwide non-military requirements of the market in the 1920s – good design, jewels, beauty, watches as fashion accessories, and so on. Audemars Piguet had been early into the market; in 1909 they made a wristwatch with a minute repeater which was sold in 1925 to Metric Watch Company, New York, then their American agent. Its design is elegant and simple – a cask-shaped face on which no name or logo appears, only bold Arabic numerals, and the hour and minute hands.

Early on, Audemars Piguet adopted the policy of delivering quality watches and complete movements to internationally renowned houses, who could sell them under their own names. Audemars Piguet's name does not appear on the face and, sometimes, not even on the movement. Among the prestigious companies for whom the Swiss firm produced watches in the first three decades of the century were Van Cleef & Arpels of Paris and Tiffany of New York. Novice collectors should be aware that some of the early wristwatches bearing these names may be, in fact, made by Audemars Piguet. Checking the movement number and Audemars Piguet's record book can confirm this.

From the beginning of the century until the 1930s, finished watches were rarely imported into the USA. Audemars Piguet, along with other Swiss watchmakers, provided only movements, usually equipped with dials and hands. These were to be inserted in cases actually made by the importer – the reason being to avoid the very high American customs duties on gold and finished watches. To maintain quality, Audemars Piguet always provided a detailed plan of the case, so that there would be no difference between the finished watch and the original model.

Another Audemars Piguet speciality also dates from just prior to the First World War. From about 1910 onwards, the company, in keeping with current fashion, concentrated on a range of men's and ladies' wristwatches, which appealed to the

The Automatic Royal Oak, with day, date and moon phases.

tastes of their wealthy clientele. One striking, if not to say extravagant, example from these early days was a small ladies' wristwatch, made in 1911. It was richly set in diamonds, with a minute repeater and central second hand; the lugs were also diamond-studded. Similar but even more spectacular was a ladies' wristwatch produced in 1920. This was placed inside a case set with diamonds and measured barely 18mm in diameter. The watch's minute repeater worked by pressing on the diamonds of the case set at 6 o'clock. The owner of this miniature marvel could also remove the watch and insert it into the pocket watch case supplied with it.

OPPOSITE *Audemars Piguet's 18 karat gold Automatic. A classically simple watch, with a white enamel dial set off by its black leather strap.*

Audemars Piguet watches like this now fetch fabulous prices when they turn up in salerooms, but, during this period, Audemars Piguet also produced high-quality watches of more sober design. Catalogs and advertisements show simple watches with top-quality movements, as well as watches with special features, such as digital display with alternate hours, or various complications such as calendars, chronographs and striking mechanisms. Men's watches generally came in square, rectangular or cask-shaped cases, the white faces featuring mainly boldly legible Arabic numerals. In design terms, these are timeless classics, as functional and stylish now as they were then. Eminently wearable, with no hint of vulgarity, these are aristocrats among antique watches and very much in demand among those with taste as well as money.

Another charming range from this period were ladies' sports wristwatches. Designed for daytime, casual use, as opposed to formal wear, these are something of a misnomer, since they are not 'sports' watches in the modern sense with chronograph, shock-resistant qualities and so on – although, no doubt, ladies went motoring or played golf while wearing them. Usually in slender rectangular cases, these are undeniably feminine and decidedly chic in the then fashionable Art Deco style. Linear or geometric patterns are inset in the case in shimmering, often brightly-colored enamel, while, on some pieces, part of the face itself is enameled. Small and slim, these were made to look good on the wrist, their appearance enhanced by the soft suede or lizard-skin straps on which they were often mounted. Like the plain, elegant men's watches of the period, these are practical, wearable timepieces today and as much sought after because of the revival of interest in 1930s styles as for their high-quality movements.

An Audemars Piguet lady's watch sold at auction by Sotheby's on February 25, 1988. The numerals are rectangular sapphires.

At this time, of course, other illustrious Swiss watchmakers were producing wristwatches, seemingly as attractive and as well designed as Audemars Piguet's watches, at a somewhat lower price, even if by no means cheap. What made Audemars Piguet's watches so special and so expensive? The reason lies in the nature of the initial philosophy of the company's two founders.

Early in their partnership Audemars and Piguet adopted the practice – revolutionary at the turn of the century – of controling the manufacture of their watches, piece by piece, from start to finish. The majority of watchmakers, then as now, 'assembled' watches from externally supplied parts. Although Audemars Piguet's suppliers were invariably highly skilled craftsmen, the partners realized that the only way they could maintain absolute quality was to bring together all such craftsmen under one roof, supervizing each individually handmade piece, finishing and testing it in the Le Brassus workshop. To this day, Audemars Piguet ignores mass production. From the first roughing out of the basic movement to the ultimate finish, each watch is the work of a master watchmaker who can take from six months to a year to produce just one watch. It goes without saying that only the highest quality materials are used. The quantity of gold and platinum is never stinted in the interests of economy and only the finest diamonds and precious stones are incorporated (each watch carries a certificate confirming its authenticity and quality). Because of the individual attention given to each watch, no two watches are exactly alike, even if they are of the same design.

A classic Audemars Piguet 18 karat gold wristwatch sold at auction by Christie's, London in October 1988.

Audemars Piguet's main claim to fame, however, lies not so much in the supreme quality of its watches but in the technical innovations it has pioneered from the early years of the century to the present day. Indeed, this company has won more gold medals in the Olympics of watchmaking than any of its rivals. Although its early firsts relate to pocket watches, the vigorous striving for technical perfection is obvious. The Grande Complication, created in 1915, required the assembly of some 400 pieces, is still made today and is one of the most expensive non-custom-made pocket watches in the world; in 1925 Audemars Piguet created the thinnest pocket watch in the world measuring 1.32mm; the skeleton pocket watch followed in 1934 and in 1946 the thinnest wristwatch in the world (1.64mm) was created.

During the 1970s and 1980s Audemars Piguet continued its research and development to produce the modern classics available today. These, too, were firsts in the history of watchmaking. One of the most enduring in more ways than one was the Royal Oak, the first luxury wristwatch ever created in stainless steel. Two years of work elapsed before, in 1972, technicians and stylists were able to produce this elegant sports watch, which became and has remained a highly popular classic. The company named the watch after the hollow tree trunk in which Charles II of England is said to have once sought refuge and which, since then, has come to symbolize sheltering

A gold Audemars Piguet with moonphase and three subsidiary dials. The integrated gold chain mail bracelet helps make this a very expensive wristwatch, and it highlights the brilliant white enamel dial.

strength – it is no accident that Britain's Royal Navy has christened at least three of its ships with this name over the years. Steel, the most intractable and demanding of metals, was combined with 18-karat gold in the famous octagonal design – which looks rather like a porthole – and in which the visible screws, intended to show the strength of the watch, are an integral part. The watch was an instant success and its design copied – but as Audemars Piguet say, 'never equalled' – worldwide.

Today, Royal Oak watches are available in steel, steel and gold, gold, or gold with precious metals. The Royal Oak 'dress watch', for example, has a bezel set with 32 diamonds; the magnificent 'jewelry watch' shows a lavish use of gems with its dial set with 237 diamonds and 11 rubies, and its case and bracelet set with 454 diamonds – rarely, if ever, has a sports watch displayed so much conspicuous wealth! And rarely, too, has such a dazzling brilliance been seen beneath the sea for, regardless of model, all Royal Oak watches are water-resistant to a depth of 50 meters (164 feet). Spurred on by the original Royal Oak's success, Audemars Piguet developed further refinements. The Royal Oak with day, date and moonphases was introduced in 1983 and the Royal Oak perpetual calendar in 1984.

A spectacular Audemars Piguet diamond-set 18 karat Skeleton watch. This range retails at over $45,000.

Audemars Piguet, as specialists in complicated watches since the late 19th century, created numerous perpetual calendar watches, many of them manually-wound pocket watches. In 1978, the company launched a Perpetual Calendar in the form of the automatic wristwatch, programed with such a complex mechanism that leap years are accounted for, keeping perfect time, untouched, until the year 2100. The classic design of this most elegant watch has spawned hundreds of imitations, all incorporating the attractively colored moonphase on the dial, probably its most visually arresting feature. Like other Audemars Piguet models, the Perpetual Calendar is available – at a price – as a jewelry watch. In the platinum version, the pavé dials are set with 271 diamonds and the bezel with 40 diamonds and eight sapphires; this sells for over $44,000 (in 1989). The more austere gold Perpetual Calendar has a bezel set with 80 small diamonds with a bracelet in 18-karat yellow gold and mother-of-pearl. By contrast, the remaining available model in platinum with automatic skeleton movement is unadorned and strenuously muscular.

Another first in the Audemars Piguet records was the development of the automatic tourbillon wristwatch. The tourbillon movement was invented in 1795 by Abraham-Louis Breguet, the master watchmaker renowned for his brilliant mechanical expertise. His invention was remarkably ingenious – and guaranteed perfect timekeeping in a watch. Instead of being placed separately, wheel, lever and balance are held together in a very light mobile cage. Drawn by the wheels, the cage revolves at about one turn per minute. The constant motion of the escapement assures the watch's precision.

After much research, Audemars Piguet was able to incorporate the tourbillon escapement – which is expensive and difficult to produce – into an extra-thin automatic wristwatch. The design of the 18-karat gold dial, in the form of sun rays, was inspired by the Egyptian Sun God, Amun-ra, who, according to legend, gave the world the gift of fire. Since its development in 1986, only a few numbered and pre-ordered examples of this exceptional watch have left the Le Brassus factory, so it is already on the way to becoming a rare collector's item.

Audemars Piguet's range of watches is relatively small. Among the plainer models are the Philosophique watch, made for both men and women, and the automatic Sportive. Its jewelry watches are spectacular and the Baroque and Dôme models, to name but two, are as much pieces of high-fashion jewelry as they are watches, and obvious symbols of status and wealth; the Rivière Dôme costs in the region of a staggering $193,800.

The sumptuousness of these jewel-encrusted pieces should never blind anyone to the refinement and quality of Audemars Piguet watches as instruments of precision. Unlike many of their rivals, the company is uncompromising in producing a limited and exclusive range of watches for the connoisseur. In 1986, a year which saw the production of 320 million watches, Audemars Piguet contributed just over 11,000 to the grand total; Patek Philippe and Vacheron & Constantin together produced some 22,700, while Rolex contributed 450,000 to the global figure. In Audemars Piguet's opinion, quality rather than numbers increases profitable turnover.

BAUME & MERCIER

The history of Baume & Mercier began around 1830 when the Baume family were already making watches in the Jura near Berne. Nearly a century elapsed before a member of the Baume family met a Geneva jeweler of high repute, Paul Mercier, and in 1918 the two joined forces, with their headquarters in Geneva. Their complementary skills insured success. With Baume producing movements of great precision and Mercier designing cases and dials with the artistry for which he was renowned, they soon became one of the top watchmakers of Switzerland, and in 1921 were awarded the coveted Poinçon de Genève, official recognition of the faultless quality of their products. Watches dating from this early period are now rare collectors' finds, worth tens of thousands of dollars.

Baume & Mercier continued to produce a range of high-quality watches for the next three decades, but by the 1960s, a once flourishing market had shrunk to include only Italy and some US outlets. Although Baume & Mercier were still good at making watches, they lacked the aggressive management and marketing skills that were so necessary to maintain credibility in a competitive international market.

By good fortune, they were introduced to the revitalized Piaget company by their Italian distributor – who also sold Piaget watches – and in 1964 Piaget acquired the major shareholding in the company. For Piaget the benefits were obvious. They had access to a potentially wider market (the average cost of a Baume & Mercier is about $2,550) without devaluing their own 'haute couture' image. Their management skills, moreover, meant that by the 1980s Baume & Mercier watches were sold in 70 countries across five continents. The company was allowed considerable autonomy and encouraged to develop its own brand image. This sound policy has paid off handsomely. In 1989, Baume & Mercier accounted for about 40 per cent of Piaget's turnover, with the US taking about 20 per cent of the company's output, Italy, France and Switzerland accounting for another 60 per cent, with the remainder going to the Far East, notably Japan.

RIGHT *A lady's 18 karat automatic Baume & Mercier wristwatch, with a tiger's eye dial and applied gold baton numerals and date aperture.*

The modern Baume & Mercier image is based on two distinctive watch styles – the classic and the sporting. These are exactly suited to the middle price range market, but with some 600 models Baume & Mercier constantly monitor their collections, discarding old watches and introducing new ones in deference to current trends. Despite this, fashion does not dictate to Baume & Mercier, whose styles really do possess a timeless elegance, which could be the envy of their more expensive rivals. Baume & Mercier watches are generally unpretentious timepieces of high quality designed as watches, not pieces of jewelry.

To be sure, Baume & Mercier watches are not produced in house entirely from start to finish like Piaget's. The company relies on outside suppliers to provide basic components, but all the assembling and finishing is done in the company's workshop, which employs about 50 people. Baume & Mercier produce about 100,000 watches per year, so inevitably their approach is fairly industrialized. They are watchful of modern developments as well. They were quick to recognize the potential of the new quartz watches and from 1970 phased out mechanical movements, so that by the 1980s nearly all their watches had quartz ones. Platinum, gold and precious stones are not used to excess, simply because they are rarely appropriate to the style – or price – of Baume & Mercier watches.

One of their most successful models, launched in 1980, is the Riviera, a sports watch par excellence in gold and steel with a distinctive 12-sided bezel and a bold, uncluttered face. To test its durability and precision, this watch was mounted on the wheel of a BMW M1 before the start of the Le Mans 24-hour auto race. It withstood high speeds as well as the pressure of enormous acceleration at the beginning of the race; the centrifugal force of the spinning car wheels failed to affect it, as did intermittent heavy rainfall and the intense heat of overworked disk brakes. The watch ran with as much precision after the race as it did before. But Baume & Mercier watches are like that – they do what they are meant to do perfectly.

Take the Medicus watch, for example, especially designed for doctors. Clean and clinical to look at, it includes a date indicator as well as a pulse scale to test a patient's

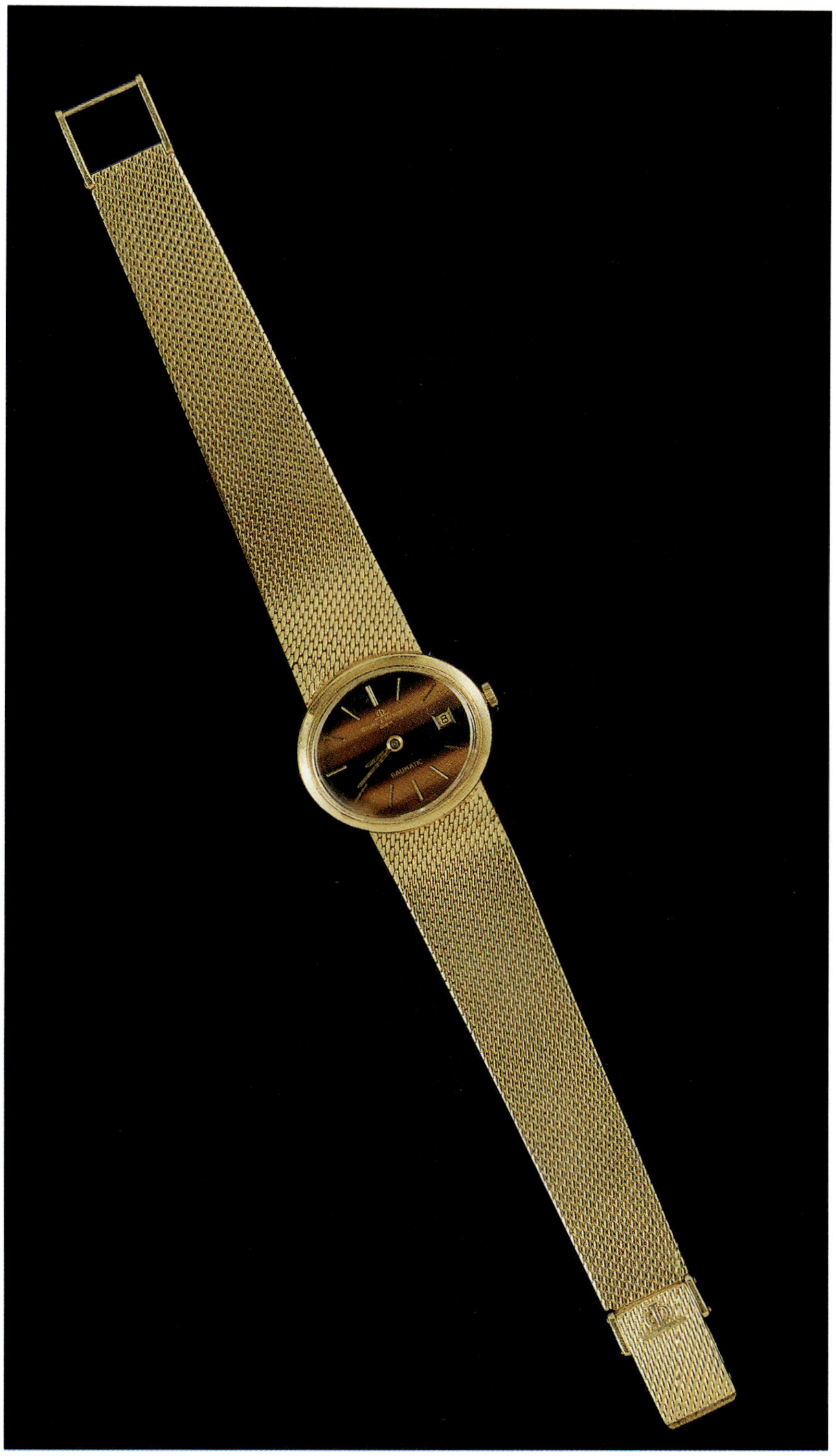

pulse rate. The Avant Garde, another sports watch, is reminiscent of Piaget's hugely successful Polo, perhaps deliberately so. Water-resistant to a substantial depth, this is even more durable than the Riviera. Bands of tungsten carbide are interspersed with strips of gold to form both the bracelet and the face. Linea, available in both ladies' and gent's sizes, is gold-plated, its case style having that faintly utilitarian quality associated with the 1940s and 1950s; a chunky wedge-linked bracelet cleverly offsets this. This is an example of the trend towards nostalgic designs, fashionable in the late 1980s.

Among Baume & Mercier's deluxe range are the Haute Joaillerie models. These do display dazzling gems, but, unlike Piaget's jeweled watches, they are stylish and chic in a fairly subdued way. Even with these top of the market models, Baume & Mercier still make watches that are superbly functional timepieces – and above all look like watches.

BERTOLUCCI

Remo Bertolucci has jumped out of an airplane more than 1,500 times, and it is the 40 seconds or so of free-falling, before he pulls the parachute cord, that he really enjoys: this proves he has no nerves and can count, and is therefore clearly qualified to be a maker of wristwatches. He is also a very good one.

Remo is a lucky Tuscan, brought up in Pisa and trained in electro-technical engineering. At 14 he was on a skiing holiday in Grindelwald, where he met a pretty girl called Pierrette Michelotti. Her father had bought a small watchmaking business in 1939 at Evilard, above Biel; its original founder in 1911 was Robert Chopard. He married her, acquired Swiss nationality, and soon took over the family business. Today the firm bears his own name and has some 35 employees, who are responsible for a small but impressive range of classic quartz and automatic wristwatches. A distinctive feature of Bertolucci's Pulchra (the Latin for beautiful) collection is the way in which the gold (18 karat), steel or mixed bracelets are an integral part of the watch design. Top of the range is a ladies' pavé 18 karat gold watch and bracelet, cobbled with 1,105 diamonds. It is restrained and elegant, the ostentation of the diamonds is minimal because of their even distribution around the wrist, while the diamond numberless dial, with no second hand or date indicator, is correctly understated. This $85,000 watch is a beauty.

BELOW AND OPPOSITE
Wristwatches from Bertolucci's Pulchra range in 18 karat gold or polished steel and gold combined. The use of rounded pebbles was an eye-catching image for this highly original watchmaker.

TACHYMETRE
DIM 14

BLANCPAIN

The story of Blancpain is the longest and probably the most remarkable in the history of watchmaking. Some of their dials feature the initials I.B. and the date 1735; with these the whole, moving story begins.

Imer Blancpain was christened on 13 May 1639 in the village of Villeret, beside the river Suze in the Swiss Jura valley of Erguel. Parts for watches had already started to be manufactured in the area, and Jehan-Jacques Blancpain (born 1693), great-grandson of Imer Blancpain, started the family business, making ébauches and parts in 1735, according to the earliest records. He put workshops on the first floor of his 100-year-old farmhouse, right beside the banks of La Suze. His son Isaac concentrated on being the Mayor of Villeret, but one of his grandsons, David-Louis (born 1765), began exporting the family's goods, which was an extremely hazardous occupation in those revolutionary days. It was profitable enough to allow the business to begin expanding on a very modest scale. David-Louis had five sons, and one of them, Frédéric-Louis, initiated the manufacture of complete watches in 1815. Just 15 years after this, his son Frédéric-Emile (born 1811) took over the firm, introducing serial production – a new factory was opened in about 1860 – and, in 1869, watches with crowns, instead of cumbersome keys. Again, a son, Jules-Emile (1832–1928) succeeded, followed by a grandson, Frédéric-Emile (1863–1932). Frédéric-Emile Blancpain was the great-great-great-great-great-great-great grandson of Imer Blancpain, and sadly he was the last of his line.

This contemporary Fortis Harwood wristwatch reflects John Harwood's original 1924 invention of a self-winding mechanism. No crown is therefore necessary.

Jules-Emile and Frédéric-Emile made many innovations including lever and cylinder watches; among them was the introduction of a 3¾ ligne baguette movement, which led to the manufacture of wristwatches. By 1926, after a late entrance into the wristwatch field, Blancpain contributed to history by making a prototype of the first wristwatch to have an automatic winding mechanism; it was powered by the wearer's movements, and was the invention of John Harwood, a brilliant English horologist. In 1929 Blancpain produced the watch for the French market. This truly important contribution to the craft of wristwatch manufacture soon produced variations, and, starting in 1930, the company made the Rolls, the dial design of which recalled the famous radiator shape of Rolls-Royce motor cars. This watch was a pun: the movement was mounted on ball-bearings inside the case, and, as the movement 'swung' back and forth or 'rolled', powered by the wearer, so the winding mechanism operated. This system had been invented by Léon Hatot, a Paris horologist.

A gold and steel automatic calendar watch by Blancpain. It has a gold bezel and winder, raised gold Roman numerals, an inner date ring, day and month apertures and moonphase at six o'clock.

The company was now dedicated almost exclusively to the production of wristwatches. They were produced under the supervision of the redoubtable Madame Fichter, who took over the running of the company after the death of Frédéric-Emile Blancpain in 1932, and continued to run it for nearly 40 years. It traded under the name Rayville SA, succ. de Blancpain; Rayville is a single phonetic anagram of Villeret, where the firm's first workshop was located. The production of the Harwood automatic and the Rolls continued, and today, depending on their condition, they are collectors' items. The Second World War came and went; the next noteworthy watch was the Air Command (1951); it was a chronograph with a 30-minute timer and a steel case with a movable glass (collectors should note that only about 1,000 were made). The first great postwar success was the Fifty Fathoms (1953); its excellence and water-resistance (200m/656ft) were confirmed by Jacques Cousteau and his crew who wore these watches during the filming of 'Silent World' (1956). Successors were Bathyscope, which was the name of Cousteau's diving chamber, and the Fifty Fathoms Milospec; the latter had a hole centered above the baton six, which indicated when damp had affected the mechanism. In the year that 'Silent World' was astonishing film and television audiences around the world, Rayville launched a notable ladies' watch, called Lady Bird; it incorporated the world's smallest ladies' automatic movement (5 lignes/11.85mm). After 1959 only wristwatches were produced.

It has been thoroughly documented elsewhere that the beginning of the 1970s witnessed a decimation of the Swiss watch manufacturing industry, because of the huge inroads made into their markets by Japanese quartz watches which caught the Swiss unprepared. Many of the old firms stopped trading, some voluntarily merged

WED
NOV

to form large, stronger groups, whilst others were faced with being taken over at more or less any price they could obtain: one of the latter was Rayville. In 1970 the company was acquired by SSIH (known as SMH today), primarily because the big combine wanted to acquire the little firm's centuries-old 'tricks-of-the-trade', its accumulated knowledge, experience and tooling. At this time Rayville and Blancpain ceased to appear on the dials of wristwatches; and the old company was put, not to death, but to sleep.

Then a happy event occurred. Another of the companies taken over by SSIH was Omega, and several years later its managing director, Jean-Claude Biver, became fascinated by the history and achievements of the ancient, dormant associate company. So he bought it. By January 9, 1983 the name of Blancpain was once again in independent use. Biver had previously been sales manager for Audemars Piguet; now he invited his friend Jacques Piguet (whose ancestor was Louis-Elisée Piguet) to join him in his new venture.

TOP *Blancpain's elegant minute repeater. Very few of these watches are released on to the market each year.*

BOTTOM *An automatic Perpetual Calendar watch by Blancpain in stainless steel, with a moonphase aperture.*

Today Blancpain is most deliberately old-fashioned, and is unique within the wristwatch manufacturing industry for several reasons. The factory is installed in the very farmhouse in which Louis-Elisée Piguet made his first movement in the watchmaking village of Le Brassus. Employees grow flowers, fruit and vegetables for themselves in the garden. Among the 30 or so employees, there are 15 watchmakers who really do assemble, polish and finish all the parts which go into a watch themselves; then the movement and case are numbered and signed for in the register. The key to the whole operation is simplicity. The total output is about 6,000 watches a year; workers work whenever they please; there is only one shape and basic design of case (round) and only two sizes (gentlemen's and ladies'). They have never used a quartz movement and never will. Of course, there are choices of dials (always on white with roman numerals) and metals, but now a Blancpain watch is always recognizable: a handmade classic, of limited quantity, and always traceable to that register and therefore to its maker. In what other major industry, involving mechanics, is it possible to meet (and Blancpain encourage you to do so) the man who made the whole of what you now own?

In 1984, Blancpain introduced a unique ladies' 8¾ ligne watch; it revived the moonphase calendar, showing the month, the date, the day, and the moonphase. A year later they brought out their first perpetual calendar, in gold and steel. These innovative models were part of a deliberate manufacturing and marketing philosophy which had been proudly and conscientiously evolved by Jean-Claude Biver. For this reason, it is relevant to repeat here the six great masterpieces in wristwatch-making. These are: Chronometer; Minute repeater; Moonphase calendar; Perpetual calendar; Tourbillon regulator; Ultra-thin chronograph.

Blancpain now produces all six of these great watches, and is determined ultimately to combine them all in one case; it will be one of the finest and most collectable wristwatches of all time. Blancpain watches are distinguishable by the movements (listed above) and materials they incorporate: 18 karat gold, which may be pink, white or yellow, platinum and 18 karat gold with steel, with or without precious stones, metal bracelets and hand-sewn leather straps (which are interchangeable). The watch glasses are all scratch-resistant sapphire, and all dial indicators are in 18 karat gold.

Here is a description of Blancpain's minute repeater watch, of which only about ten leave their farmhouse workshop each year: a summary of its ingredients certainly proclaims a classic wristwatch. The ultra-thin self-winding movement, with a central rotor in 22 karat gold, is 3.2mm thick and 21mm in diameter; it uses more than 30 rubies and has more than 300 parts (weighing less than three grams together) and some of them are thinner than a human hair (which is 0.06mm in diameter). This watch can repeat, whenever required, the hour, the quarter hour and minute, in a combination of chimes (by two hammers, each with its own tone and vibration), without hindering the main recording timepiece. On the side of the case beneath the bezel at nine o'clock, there is a slim loading lever which releases the sound of the chimes when it is pushed towards 12 o'clock. It is a sort of signature tune, confirming the continuing existence of a long line of masterpieces which rank among the most desirable ever made; little heirlooms which are both beautiful and useful.

AUDEMARS PIGUET
AUTOMATIC

BREGUET

Breguet's gold automatic moonphase and date wristwatch. Note the unusual engine-turned silvered dial with its Roman numerals and blued steel moon hands, date ring at six, a sector at one with moonphase and lunar calendar, and a sector at ten-thirty, with a power reserve indicator.

'Time is the greatest innovator', wrote the 17th-century English philosopher, Francis Bacon. Abraham-Louis Breguet (1747–1823) was one of the few whose innovatory genius has had an impact on time, or more precisely its measurement. So extraordinary was the mechanical ingenuity of this Swiss-born watchmaker that he gained and maintained the patronage of royalty, the rich and the powerful throughout Europe at one of the most turbulent times in its history.

Breguet thus found himself in the uncommon position of providing watches for the *ancien régime* and later the power brokers of the new revolutionary France; in 1815 both Napoleon and Wellington were consulting their Breguets at Waterloo. A Breguet watch accompanied Alexander von Humboldt to the New World and was the preferred timepiece of the tsars of Russia. Indeed so legendary did these horological masterpieces become that they were immortalized in fiction. Dumas' Count of Monte Cristo sported one, as did Phileas Fogg, the intrepid and very time-conscious hero of Jules Verne's *Around the World in Eighty Days.*

The master watchmaker himself was of French Huguenot descent, whose family had been forced to flee Paris as a result of Catholic persecution. After Abraham-Louis' father died in 1758, his new stepfather, Joseph Tattet of Verriers (a successful watchmaking business based in Neuchâtel and Paris) was quick to note the young man's talent. At the age of 15 Breguet was sent to Versailles to become a watchmaker's apprentice.

During his five years' apprenticeship, Breguet attended evening classes in mathematics at the Collège Mazarin, an essential background and training for a man who was intending to distinguish himself in the production of precision instruments. Through a series of fortunate incidents, he soon came to the notice of Louis XV, an auspicious beginning to a career at a time when royal patronage was important for success.

Breguet married in 1775, and his dowry permitted him to set up both home and business in the prestigious Quai de L'Horloge on the Ile de la Cité, the heart of Paris. It is ironic that only 18 years later his first great patron, Marie Antoinette, was to spend her last days on this same quai, incarcerated in the grim 14th-century prison of the Conciergerie.

Breguet's carefully kept registers are still preserved. Dating from 1787, each Breguet watch is recorded with the name of the individual maker, the cost price, sale date and identity of the purchaser. The name of the French queen appears frequently in the first few pages. With the characteristic extravagance that was finally to prove her undoing, she is recorded as buying her pocket watches in batches of six, a fashion which was imitated by the court. Breguet was already at the top of his profession.

It is to Breguet that we owe the first automatic watch (the perpetuelle). The principles of the self-winding watch were probably first invented and unsuccessfully put into practice in 1765, but it was Breguet who undoubtedly brought the perpetuelle effectively into existence. According to extant writings of the watchmaker, Marie Antoinette and the Duc d'Orléans each possessed such watches by 1780, no doubt inscribed, along with the earliest examples, 'Inventé et Perfectionné par Breguet à Paris'.

Breguet's genius was to transform the basic but inefficient self-winding mechanism of A. L. Perrelet into a sophisticated machine. Two barrels, connected to a platinum weight pivoted on an edge of the backplate, were constructed into the watch mechanism, so as to derive maximum response from every movement of the wearer. Four turns of the barrels powered the equivalent of 60 turns of the center wheel (less than two turns can run the watch for a day).

It has been estimated that a mile or so of ordinary walking will wind the perpetuelle sufficient for 60 hours' operation. The earliest extant Breguet self-winding pocket watch dates from October 1783, though there are records for a similar model sold to Marie Antoinette for 4,000 francs exactly a year previously. A special feature of these models was a fan-shaped hand and scale ranging from zero to 60 on the dial to indicate the number of hours' running time left to the user. Another trademark

An 18 karat yellow gold Breguet wristwatch, which was auctioned by Christie's, London, on October 28, 1988. The watch number is 347, and the signed, plated and polished movement is numbered 521. The fine open-weave 18 karat Breguet strap and clasp no doubt contributed to the pre-sale estimate of $3,000–$4,500.

introduced by Breguet, and universally imitated, was the tiny circular moon on the hour and minute hands. Both appear on Breguet watches to this day.

In the meantime Breguet financed his experiments by importing complete watches and ébauches from Switzerland, which were finished to his exacting artistic and mechanical standards. These inevitably fall a little short of the extraordinary quality of those he had custom-built for his wealthy clients. The intellectual excitement of applying his invention to the subtle problems of horological engineering was the breath of life to Breguet.

The perpetuelle was being manufactured in quantity by 1786, and Breguet entered into a partnership to raise sufficient capital to finance his expansion. His association with Xavier Gide was of six years' duration, serving to put the business onto a firm financial footing; it marks the beginning of his records. Some of the perpetuelles produced at this period feature for the first time a new Breguet invention, *Le parachute pour le balancier*: shock-proof jeweling that protected the delicate watch movement from the damaging effect of being accidentally dropped.

Many of Breguet's pocket watches were repeaters, sounding the hours, quarters and half-quarters. More complicated versions also registered each ten minutes, five minutes or minutes with a cunning series of distinguishing blows, testimony to Breguet's fascination with the solution of highly complex horological engineering problems.

By 1793 Breguet's position as the leading Parisian watchmaker had become increasingly perilous. The majority of his aristocratic connections had either fled the Terror or, like Marie Antoinette, had ended their lives at the guillotine. That August Breguet escaped to Switzerland, where he was to remain for nearly two years. This difficult time was nevertheless a highly creative one. It was there that he conceived the perpetual calendar, *la montre à tact*, the *souscription* or one-hand watch, *La pendule sympathique* and, most important of all, the *tourbillon* or rotating carriage watch. The patent is dated 1801 but needless to say there was a long period in which Breguet was perfecting his invention.

La montre à tact was, like the repeater, an invention for determining the time in the dark in the days before luminous dials. An arrow, which could be ornamented with precious stones, was set into the bezel and the time was ascertained relative to touch pieces made of diamonds, pearls or other material. Breguet's sympathetic clock was another invention which did not survive. A specially designed watch was positioned into a table-clock with a half-moon fork and overnight both wound and set to the exact time registered by the clock. It was also during this period of exile that Breguet began to use his secret signature. The method of inscription which utilized a small pantograph was devised by his friend the medalist Jean-Pierre Droz. The result (the name of Breguet and the watch's individual coding with the addition of *'Souscription'* if of that variety) is so tiny that it can only be read with the aid of a magnifying-glass.

Breguet was not however allowed to stay away from France for long. His talents were demanded for the reorganization of the Versailles watchmaking center and the equipping of the army and navy with advanced horological instruments. Both his house and workshops were restored, and Breguet continued to produce his marvelous watches, each taking months to complete by a master-craftsman (a repeating watch could take up to twelve months; two years for a perpetuelle). About four thousand clocks and pocket watches have been manufactured between 1794 and 1823.

Perhaps the acme of his craft was the celebrated complex watch known as the Marie Antoinette. Ordered in 1793 by an officer of the guard of that unfortunate queen, it was requested that it should include all the complications then known, regardless of expense and time. The parts usually reserved for brass were manufactured from gold and could be seen through the rock crystal dial and backplate. It included the perpetuelle winding mechanism and indicator, a time equation, perpetual calendar with day of the week and date indicators, thermometer, and repetition for hours, quarters and minutes. Completed in 1820, this marvel of the 18th-century horologist's art had cost a total of 16,484 francs and was finally kept by the master himself, a fitting tribute to his lifetime's work. Tragically it disappeared from sight after a break-in at the L. A. Mayer Memorial Institute in Jerusalem, and today its whereabouts are unknown.

The house of Breguet has continued producing watches to this day, though the business passed out of the hands of a member of the family in 1870. Exactly a century later it was owned for a time by the Parisian jewelers Jacques and Pierre Chaumet, who were determined to revive the flagging reputation of this once great business. Recently Breguet changed hands again.

A workshop was set up in Le Brassus in the Swiss Vallée de Joux staffed with watchmaking craftsmen of the highest caliber. Each watch is traditionally made by hand, often using the methods and tools of two centuries ago. Breguet watches were never merely finely tuned precision instruments. They also had to satisfy the artistic sensibilities of sophisticated clients, a task they still manage today, in an era when the wristwatch has all but superseded the pocket watch.

The style of the modern Breguet wristwatches echoes the pocket watches of Breguet's best period. The dials are still engine-turned by hand. The elegant milling on the silver-plated face is distinctive as is that which also traditionally ornaments the edges of the 18 karat gold watch case. Jewels are used with taste and flair on the case, bracelet or lugs of some examples, in contrast to some of the more vulgar displays of modern watchmakers. Skeleton watches, whose visible movements are set with diamonds and rubies, are a triumph of the jeweler's art.

The modern Breguet wristwatch – many of them extra-flat – offers a variety of options. Indicators for the date, phases of the moon, variable second hands, perpetual calendar with leap year indication, are the basic mechanical possibilities, set off by 18 karat gold woven chain bracelets or leather straps. All bear the name Breguet and each carries its unique production number on the dial, a homage to their inspirer and a testament to a continuing tradition of unique craftsmanship.

BREITLING
NAVITIMER

BREITLING

Breitling exhibits each year at the world's biggest airshow in Wisconsin; to realize why is to recognize and understand the trading philosophy of this impressive independent company. The Breitling watch owner is very likely to be an active person – in outdoor sport, sailing on the world's oceans, diving beneath them, or flying above them.

Léon Breitling first opened a workshop in La Chaux-de-Fonds in 1884, making pocket watches and chronographs. His son Gaston initiated the production of wrist-watches in 1914 to provide for the wartime necessity for synchronized military action: these early models incorporated a stop-watch and had a luminous dial and hands. In 1936, Willy Breitling, Léon's grandson, launched a chronometer for instrument panels in aircraft cockpits, and the firm has been supplying them ever since, to customers such as Boeing, Douglas and Lockheed.

In 1952 a logical manufacturing extension to this close association started; this produced the famous Navitimer, a superchronograph designed specially for pilots, and used for preparing flights, checking flight plans and for calculating speeds and fuel consumption. This sturdy mechanical watch is very collectable, and is much preferred to Breitling's GMT in quartz, in different casings, introduced in 1983. The latter, though, was a novelty at the time, simultaneously showing the time zones; there are two analog and one digital versions with a chronograph; it also has two separate movements, with independent sources of power, the value of which pilots obviously recognize.

The first Navitimer went supersonic in 1962; a 24-hour display quartz version of it was launched with the name Cosmonaute, and the astronaut Scott Carpenter was in turn launched with this watch somewhere near his wrist. For less ambitious supersonic air-travelers, Concorde regulars and customers with international business interests, Breitling introduced in 1970 the Breitling GMT, a chronograph with not one, but two, hour hands (one on a 12 hour dial and the other showing 24 hours). A year later the company surely anticipated the needs of even the busiest tycoon: the Breitling Unitime simultaneously shows the time in all the countries of the world.

Breitling customers are demanding, and, since Ernest Schneider (a qualified pilot and electronics expert) acquired the company in 1979, he has been most careful to continue its long tradition of observing minutely the requirements of a particular market sector. A classic wristwatch may or may not be very beautiful, but it must always be useful – compatible with its wearer's needs.

Breitling's quartz steel Deep Sea (1985) is such a model, built to withstand underwater pressure at 1,000m/3,280ft (a 10 times safety factor). A gloved diver can record different dive times with the one-direction click-stop bezel timer, and he can read the dial at any depth (the minute hand is much wider than the hour hand, because diving times are naturally short). Divers at great depth often fear the battery running out:

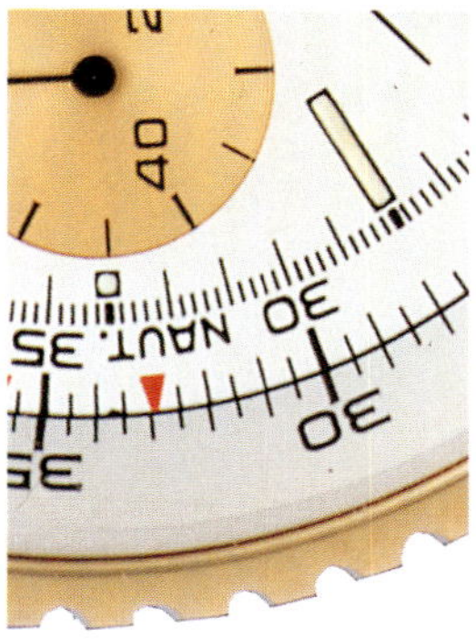

FAR LEFT *This Old Navitimer is a good example of Breitling's craftsmanship and has been in constant demand, in its various models, since 1952. Not only does it offer chronograph timing in hours, minutes and seconds, but also a slide-rule on the outside of the face; this is most useful for pilots' in-flight calculations (for example, how much fuel remains in relation to the miles travelled, at what speed in relation to the journey time should the aircraft be flying, etc). The face is available in 12 and 24 hour displays.*

LEFT *A detail of the dial of the Old Navitimer.*

BREITLING
NAVITIMER

LEFT *Breitling's titanium Aerospace, with both analog and digital displays. This multi-functional watch has color dials which are treated against the influence of UV rays.*

LEFT *The Breitling World has four time zones, each with its own independent quartz movement. It can also be used as a solar compass.*

the second hand on the Deep Sea progresses only at four-second intervals when its battery begins to run down, thus conserving power and alerting the wearer. This watch also has an unusual patented feature. Helium gas can permeate anything, including a steel watchcase; the Deep Sea would explode on arrival at the decompression stage after a deep dive, if it was not for the twin release valves set into the bottom of the case; they open fractionally for a few thousands of a second to release accumulated helium, without letting water in.

A watch that can save lives? The Breitling Emergency can, and James Bond fans and other adventurers will greatly approve. The prototype of this amazing watch was introduced at the Basle Fair in 1988, but Ernest Schneider, ever conscious of his alpine military experience, was determined to perfect the piece in terms of battery life, and full production has been slightly delayed. Essentially, the lower part of the case contains a miniature transmitter with an activating antenna, which has a range of between 3 miles and 12 miles, depending on the nature of the surrounding terrain. Its unmodulated AM signal will transmit at two pulses per second, uninterrupted, for between 20 and 28 days, at temperatures between −20°C/−4°F and +100°C/+212°F. The transmitter, which is watertight and works underwater, can be switched on manually, or automatically when the antenna is extended. Additional features include a highly polished watch back, engraved with international air/ground symbols, which can be used as a sunlight reflector, tritium dial and hands for increased legibility, and a silicone rubber bracelet with clever parallel halves, that can be used for map reading.

CARTIER

ABOVE *A selection of Cartier wristwatches auctioned by Sotheby's, London, on July 24, 1986. The right-hand watch is an unusual 18 karat reversible model which fetched £2,530 (c. $4,300) including the 10% buyer's premium.*

BELOW *A small 18 karat mechanical Cartier Tonneau.*

RIGHT: *This is a rare 18 karat Cartier eight-day rectangular wristwatch which fetched $27,000 at Christie's, London, on March 3, 1989. (Note that the minute hand is missing and the general condition is not good.)*

Remove the letter I from Cartier, and a very ordinary English surname remains; replace the I, and once again you have an instantly recognizable synonym for an object of beauty. Cartier is one of the most famous brand names in the luxury goods market of the world; venerable, respected, admired, hugely stylish, expensive, exclusive, exciting – it possesses all these attributes and many more. How strange then that Cartier has also produced what is perhaps the most famous wristwatch ever manufactured – and also in great numbers, at a comparatively low retail price, and with an apparently simple dial and case design, black and white and all squared off. It is known as the tank watch and, due to its worldwide reputation, it is probably the most faked wristwatch. The tank made its first appearance in 1917, but the history of Cartier's association with the design, manufacture and retailing of wristwatches goes back to the last century and to the fabulous traditions of Cartier's jewelry.

It is on record that in 1888 Cartier had for sale ladies' wristwatches, with diamond and gold bracelets; these were probably the earliest fashion wristwatches manufactured for general sale, instead of being specifically ordered by a customer beforehand. It seems that they did not appeal to the passing public; a few more were made in 1892 and 1894 and took several years to sell. In fact, however, two events were occurring at this time which, in wristwatch-making terms, were momentous. First the military one: as early as 1880 Girard-Perregaux supplied a sturdy utilitarian wristwatch to officers of the Imperial Austrian Navy; it is known that in Germany in 1902 no less than 93,000 wristwatches were sold. Now for the second event – this was the year when the dictates of fashion no longer required the grandly dressed ladies of society to wear long sleeves or the previously obligatory long gloves at soirées. The coincidence of the needs of the military and the suddenly available feminine wrist led the old established firm of Cartier to ally the watchform to the bracelet. Cartier's earlier watches are heavily jeweled, definitely feminine, most beautifully designed, and expensive.

Cartier

CARTIER

More than this, Cartier, as a company, had just hit a pivotal point. The first modern Olympic Games were held in 1896; local, national and international sports activities were increasing at the turn of the century; motor cars were speeding along newly paved roads; the ultimate in trans-Atlantic liners were under construction; canals, dams, tunnels and bridges were being planned; and early flying machines were taking to the air. Outdoor life was becoming more practical, more generally shared, more fun. In the skies over Paris in the middle of the dizzy first decade of the 20th century, one man was thrilling crowds below, while he demonstrated that all these changes had occurred. He was a balloonist and an early aviator, and his name was Alberto Santos-Dumont (1873–1932). This fearless and wealthy Brazilian (Santos is the leading coffee port in the world, south of Rio de Janeiro) had a talented and wealthy friend, whose name was Louis-François Cartier. Stories of meetings of men and their subsequent collaborations are a happy and constant feature of this book; the Dumont–Cartier friendship was to bring about a pivotal event in the history of wristwatches.

FAR LEFT: *An 18 karat white gold wristwatch by Cartier, numbered 16948, with a movement by The European Watch & Clock Co.*

LEFT *A 1930s Cartier tank watch with a cabochon winder.*

LEFT: *Cartier's small Tortue in 18 karat gold with a quartz movement, elegantly set off by its serpent bracelet.*

Louis-François Cartier clearly had vision, leadership and the essential added quality of perception. When his friend Alberto Santos-Dumont asked him to produce a watch to attach to his wrist, instead of to a chain, he immediately foresaw the eventual market for such a convenient object. His balloonist friend in his basket would have both hands free to manipulate the controls, and it should, for this elegant man, be a handsome and well designed object as well, but not at all feminine. By 1904 the first Santos-Dumont was made, either for pendants or wrists, in a wide variety of beautiful designs, often heavily jeweled. In 1910 Cartier invented the deployant buckle. In 1911 a Santos wristwatch went on general sale, and it has remained available ever since: its bezel is square, but with rounded corners, it has eight holding screws, and its simplicity is unmistakably classic. The golden years of mechanical wristwatches had now begun; they were to end with the Second World War.

In 1912 Cartier brought out two more memorable designs – the oval Baignoire (French for bath or bath tub), with a rounded or rolled bezel, just like a bath edge of the period, and the Tortue (tortoise, the shell of which is recalled in the watch shape). These watches, together with the Tonneau (1906; French for barrel or cask), are all still sold, with variations and only slight modifications. So, too, is the most famous Cartier wristwatch of all – the tank. Louis-Joseph Cartier had the inspiration for this design in 1917, which he derived from the appearance of a First World War army tank with its twin tracks longer than the body of the vehicle between them; the tank watch went on sale in 1919 and has been sold (and illegally copied) ever since.

The next great watch to come from Cartier was the Pasha (1932); this was the world's first luxury water-resistant wristwatch (the Rolex Oyster was already on the market in an inexpensive model), and it was initially designed for the Pasha of

CARTIER

LEFT *Cartier's 18 karat automatic Santos.*

RIGHT *Cartier's contemporary Panthère range is in 18 karat gold, has a quartz movement and is water-resistant. Bracelets come in gold or gold and steel*

Marrakesh, so that he would know when to get out of his swimming pool. A year later came the Vendôme, inspired by Cartier's observation (with Ernest Hemingway, it is said) of a horse's harness and shaft attachments with distinctive single lugs, in Paris's Place Vendôme. The stirrup-shaped Calandre, with double lugs below and a simple one above, was another beautiful design innovation.

Cartier maintain records of every watch they have ever sold, which means that all collectors can establish the provenance of their watches. Cartier ébauches have been supplied over the years by many of the great makers and are so marked. The European Watch and Clock Co. of New York supplied many beautiful cases for Cartier, and also acted as their importers from 1919 to the mid-1950s; EWC can be found on cases. Today, the wristwatch catalogs of Cartier and Must de Cartier feature all the old names and shapes, but in the modern clothes and decorations; they are worthy inheritors of some of the richest watchmaking traditions that exist. They also look ahead to new generations of customers, and several exciting entirely new models are at the design stage.

CHOPARD

The full trading name of this glamorous company is Le Petit-Fils de L. U. Chopard & Cie SA, but the Scheufele family, who now own it, must feel like true inheritors. Louis-Ulysse Chopard, himself a member of a distinguished watchmaking family, founded the firm in the town of Sonvillier, in the Swiss Jura mountains, in 1860, and in the early days he was a major supplier of pocket watches to the Swiss railways. Business expanded, and in 1920 the company moved to Geneva.

The founder's grandson, Paul-André Chopard, started to think of retirement in the early 1960s; he had heirs who were not interested in the business, and in 1963, he sold his firm to Karl Scheufele Sr. The Scheufele family had been involved in jewelry and watchmaking for one generation longer, in West Germany's Black Forest region, at Pforzheim, which had long been (and continues to be) a center of technical watchmaking excellence. The company of 12 employees started to expand again, and in 1968, the year of Paul-André Chopard's death, it was moved to Servette in north-west Geneva. In 1972, it moved again to its present modern factory in nearby Meyrin. Today, the firm of Chopard is once again a fully-fledged family company. Karl Scheufele Sr heads it, his wife, Karin, looks after the Pforzheim end, their son Karl-Friedrich is Vice-President, and their daughter, Caroline, maintains client contacts, services and promotional services (she also features attractively in some of the brochures).

The point of this brief two-family outline is to highlight one more set of historical reasons for a company's profile and image. Chopard today is an outward-looking company with a vibrant feel, a long tradition of jewelry manufacturing, making fresh products of wide appeal at the top ends of different luxury markets; those markets like precious stones on their wristwatches. A typical Chopard wristwatch is Happy Diamonds, incorporating, between the bezel and the dial, loose gold-hooped diamonds which move with the wearer's wrist motion. The patented Happy Diamonds concept was the creation of Roland Kurowski in 1976 and there is now a very wide range of

RIGHT *The simplest wristwatch in Chopard's contemporary range.*

FAR RIGHT *Two watches in Chopard's Gstaad Collection, launched in 1986; it includes a wide range of models in gold and steel, and some models are set with diamonds.*

Chopard
GENEVE
SWISS MADE

LEFT:
Chopard is renowned for its Happy Diamonds range of wristwatches and fashion accessories. The moving diamonds pictured are seven in number, for good luck.

ABOVE:
David Penney's 1988 conceptual sketch of the evolution of a Chopard wristwatch.

wristwatches available, many of them with matching cuff-links, brooches, pendants, rings and earrings. These are the jewelry watches, mostly for the ladies' market, which clearly likes the range of choice, their gleaming originality, and the idea of diamonds softly rattling around on the wrist holds a unique attraction. The most expensive, Solitaire, retails at about $192,100.

The 1980s have seen a succession of new Chopard wristwatches, all of which demonstrate the glamor and vigor of the company's approach to different markets. Like many watch manufacturers during this decade, Chopard has strongly identified individual ranges with particular sporting events and venues. In 1980 came the St. Moritz, their first sports watch, which has four distinctive sets of twin screws around the bezel. The Gstaad collection (1986) was followed, in 1988, by the handsome Mille Miglia watch; this commemorates the re-starting of the world-famous Mille Miglia, the road sports-car race held from 1927 to 1957. In 1987, 292 of some of the greatest sports-cars ever made once again raced the 1,000 miles from Brescia to Rome and back. The Mille Miglia naturally incorporates a tachometer, luminous baton numerals and hands on a plain white dial, an 18 karat gold or stainless steel water-resistant case (to 30m/100ft), and a non-reflecting, shockproof, scratchproof sapphire glass.

The annual watch output at Chopard runs at about 20,000; 70 per cent of these have quartz movements and the rest are mechanicals. One can visualize more mechanicals coming from the Scheufele family in the near future, as their lines of contemporary classics increase and as they look again at the possibility of expanding the handsome classic ellipticals.

CORUM

1955 is unfashionably recent for the foundation of a classic watchmaking company, but there is no doubt that Corum has established, in a few years, a wide reputation for highly imaginative and innovative wristwatches. The adjectives apply to Corum's designs: these are sophisticated, fresh and thoroughly considered. Design-led they may be, but technically they are often highly competent, as in, for example, the Golden Bridge (1980). The comparative youth of the company and its management should in no sense distract the attention of wearers, collectors and investors from the virtues of Corum wristwatches.

The name derives from the Latin word 'quorum', which means the minimum number of people required (according to the relevant rules) at a meeting to validate a vote. The classic calligraphy of the word 'Corum' on the dials of the company's watches was inspired by a chiseled Latin script, and the logo of the equilateral golden key (facing upwards instead of left or right in search of a lock) was designed by René Bannwart as 'the key to perfect time'. Bannwart, and his cousin Simone Ries, started their company with her father Gaston, who had been operating a small watch factory in La Chaux-de-Fonds since 1924. To this day the firm of Corum Ries, Bannwart & Co remains a family concern in the same valley; each watch in its small output is assembled and crafted by hand, and carries its own unique serial number and certificate. For collectors who care for pedigree, here are some Corum watches to look out for from their range of 100 models.

The Chinese Hat (1960) has an exotic gold headdress surrounding the bezel; it has an elegant simplicity and is wholly unusual. The Longchamp (1957) also has a gold surround, a flat pitted disk in which the crown is imbedded; the winding finger is rolled over it. This watch set a trend with the placing of its lugs beneath the case, so that the watch appears just to sit atop the bracelet. The bracelet also goes through the bottom of the famous Golden Tube (1957), in which the movement is inserted in a

OPPOSITE *Chopard's Mille Miglia commemorates the famous road race for sports cars which was held annually between 1927 and 1957. A new form of the race was held in 1987 and this watch was launched the following year. Available in gold, gold and steel or steel, with white or black dials, the tachometer enables today's motorist to measure his car's speed precisely.*

Corum's remarkable Peary watch, with its unique dial composed of a polished slice of meteorite. No two dials therefore can ever be the same, and each watch is supplied with a certificate of origin explaining the discovery of the meteorite by the US explorer R E Peary (1856–1920). He was the first man to reach the North Pole in 1909, in the last of his six expeditions.

Admiral's Cup
CORUM

ABOVE *Corum's automatic Coin Watch is handmade from a genuine United States twenty-dollar coin. The extra-flat movement is inserted between the two faces of the Double Eagle coin.*

LEFT *Corum's Admiral's Cup, which may come to be regarded as a classic in future. Enameled nautical pennants marking the hours dress the dial. It is, naturally, water-resistant.*

horizontal tube, with the crown at the top; a tassel is sometimes found in the normal right-hand side crown position. The simple outsize Buckingham (1965), with its ten distinctive horizontal baton numbers, is a timely reminder of the bulk of early automatic movements.

The notion of a coin converted into a watch may not seem attractive, but Corum believed that an ultra-thin movement inserted between the faces of a gold coin might find a market as a wristwatch. The firm was correct in its guess, because it was wise enough to look back to the days of the 1840s Californian gold rush. It selected the Double Eagle, a $20 gold coin, which was officially approved by Congress on March 4, 1849, and decided that the dial side should be the reverse of this famous coin, showing the American coat-of-arms, supported by the bald eagle and surmounted by a scroll with the motto 'In God We Trust'. Only the thin black hands 'interrupt' this handsome and historic coin, and today collectors appreciate early versions of the quartz Coin Watch (1965). The lure and lore of gold later inspired Corum's quartz Ingot Watch (1977), in which only the hands pierce an authentic Union Bank of Switzerland ingot of 99.99 per cent pure gold, weighing (in various models) up to 15 grams (also available in platinum). Each ingot is numbered and accompanied by an official certificate from the Swiss assay office, confirming its gold content.

Corum's Romulus (1966), named after Rome's legendary founder and first king, was the first wristwatch to have the hours hand-engraved on the bezel: a typically fresh innovation which has been imitated widely ever since by other manufacturers. This ultra-thin quartz watch is water-resistant and comes in a number of different metals.

The positioning of Corum's name and its immediately identifiable logo are the only slight blemishes on the remarkable appearance of the Golden Bridge (1980), surely a future classic mechanical. The wheels are assembled in a straight line, beneath a vertical, jeweled 18 karat gold bridge; the clear white sapphire glass is faceted, which adds a depth and mystery to the movement.

Finally, there are two major ranges with distinctive dials to look out for from this highly original low-quantity manufacturer. The quartz Admiral's Cup (1982) is a marketing triumph, and (produced as it is with official permission from the Royal Ocean Racing Club, London) commemorates the famous Admiral's Cup yachting race off Cowes, in the Isle of Wight, every two years. Within its twelve-sided case, the dial features colored miniatures of the flags of the international maritime code, which is used for visual signals between boats; naturally it is water-resistant. The second notable dial is on the Météorite range (1987/88); some of the watches feature slices of one of the larger meteorites ever to fall to earth. Corum bought enough of the 34-ton 'Cape York' (in Greenland) to make only 999 dials, and no two can be the same. This variable range, showing a material more rare than gold, will one day be collectors' pieces.

ABOVE *An 18 karat gold version of Corum's Rolls-Royce model, introduced in 1976, with the famous 'Spirit of Ecstasy' mascot set into the lug. Car radiator watches make an attractive category for collecting.*

LEFT *Corum have a distinctive Platinum range, which they have developed since 1978. The square-cased watch on the left has the hours spelt out in French, in a charming period typography.*

DUNHILL

Definitions of classic wristwatches are in the Introduction and elsewhere in this book. Notice should also be taken of the words of Alfred Dunhill, founder of the now widely diversified luxury goods manufacturers and retailers bearing his name: 'It must be useful. It must work dependably. It must be beautiful. It must last. It must be the best of its kind.' Certainly his definitions of his early products ring down the years and apply equally today to the Dunhill range of wristwatches.

From the very beginning, in 1907, the Dunhill operation has been a commercially-successful exercise in marketing to gentlemen of taste, excellent design in luxury goods. Alfred Dunhill's first shop was located in London's St. James's, then, as now, the heart of clubland. Clocks and watches first arrived in his windows in 1926; the Unique lighter, which incorporated a watch in its side, is a collector's item. Two years later, with shops by now in Paris and New York as well, came the ingenious Belt watch. This was incorporated in the clasp of a trouser belt; a gentle touch on the crown pushed the mechanism downwards slightly, enabling it to spring open and discreetly reveal the time of the day to its downward-looking wearer. In 1929, Dunhill launched its first wristwatch; its plain rectangular appearance reflected the influence of the basic requirements of the First World War, but a closer look revealed the exotic, almost digital, black, Arabic numerals (the hourglass figure eight, for example). Its gold case was visibly and serially numbered on the front lower side. In the 1930s came a succession of fine gentlemen's wristwatches, featuring early luminous dials, their first chronograph, and the wonderfully original ball-race watch.

Dunhill wristwatches began their trend toward today's ranges in the 1960s, with 18 karat gold cases, and sometimes with special features, such as alarms or date calendars. Then came the round, distinctive Vermeil (1975), displaying the now familiar company logo on its gold machined dial, large roman numerals and stark black baton hands, and a crown with cabochon terminal. The single flat lugs of the Vermeil were carried over to Dunhill's quartz Millennium range (30 different styles), which was launched in 1982. This range has been highly successful and is another masterpiece of Swiss engineering, based on Dunhill's own London designs and specific requirements. It offers a choice of three sizes, round or square dials, baton, roman or diamond numerals, and various dial finishes. Dunhill have achieved a range of future classics with the Millennium; its designers have done so by laying down a number of interesting rules: baton sweep second hands only, no subsidiary dials (except on the recent Multifunction and Chronograph), no Arabic numerals, date calendar at 6 o'clock (except on model DQ

RIGHT *On the left and right are the gentleman's and lady's Millennium watch from Dunhill; in the center is a black-dial gentleman's watch with diamond-set numerals.*

OPPOSITE *Dunhill's Millennium Chronograph. Sapphire glass protects not only the front of the watch but also the back, revealing the intricate mechanical movement.*

RIGHT *Dunhill's elegant Millennium wristwatch.*

LEFT *A further selection from Dunhill's Dress Watches range.*

1728), a distinctive gold bezel within the case (which survives on the Sports models), broad single lugs, protected crowns and sapphire glass for maximum protection.

The 1986 Elite range of slim wristwatches basically adheres to the same rules, but they have ratchet lugs with more elaborate bracelets, and their dials and bezels have diamonds on some models. Among Dunhill's acclaimed Dress Watches, the 18 karat gold gentlemen's model DQ 1855 is coolly distinctive; to borrow an expression from a good wine merchant, it is a good watch to lay down. There are five dial finishes to choose from, the cabochon is a blue sapphire, the numerals are in bold roman for quick reference and the hands are black batons. Collectors and investors should take a close look at the Limited Edition Dress Watch; it features an automatic movement which is visible from the back, behind special sapphire glass. The 18 karat white gold rotor can be engraved with the customer's initials: a wonderful possession, and truly in accord with Alfred Dunhill's original dictum.

EBEL

Ebel is a modern success story *par excellence*, for the meteoric success of the Ebel range of watches over the last decade is a rags-to-riches story in the best fairy tale tradition. The hero of the story is Pierre-Alain Blum, grandson of the original founder, who has succeeded in turning around a declining firm to create a multi-million-dollar business in a little over a decade.

Blum typifies the new breed of businessman who believes in a dynamic and innovative approach to marketing. This may not endear him to the more traditional members of the watchmaking fraternity, but Blum has had to cover much ground in a very short time. No one can deny the effectiveness of his approach. By 1984 Ebel's turnover had multiplied 40 times in a decade and the workforce increased from 55 to 550 full-time employees. Ebel now holds the number three position in the luxury watches market and its ambition is to better this.

A five-year stint in Lucian Picard's watch shop in New York during his early twenties had taught Blum the value of aggressive marketing and meticulous attention to detail. When he returned to La Chaux-de-Fonds in 1972, it was in time to prevent his father fulfilling his threat to liquidate the 60-year-old family concern. A difficult partnership terminated when his father retired after an accident. Two years later Pierre-Alain Blum had bought out his father and become sole shareholder with absolute control.

Luck was on his side. The company, since its inception, had survived as a watch assembler, only rarely putting its name to the finished product. However, in 1972 Ebel had managed to secure a valuable watch manufacturing contract for the new Cartier 'Must' range at a time when other companies were suffering in the recession caused by the arrival of quartz movements. Precious time and money were gained to invest in an exclusive range of wristwatches carefully designed to appeal to the post-war generation: young, successful and upwardly mobile.

Blum was ruthless, replacing old-style management with a motivated and innovative marketing approach. By 1977 the first Ebel watch was on the market. Designed by Edy Schoepfer, the Sports line bears all the hallmarks of the Ebel image – a watch geared to the modern consumer who expects technologically high performance and durability without forgoing the refinements of aesthetics or conspicuous luxury.

In 1985 Ebel felt confident enough to launch the Beluga range. This designer watch is geared for the luxury end of the market, each example crafted from 18 karat gold, two of the models echoing the semi and perpetual calendar modes of the Sports original.

The image of the durable Sports watch has been given a high-profile sponsorship coverage, an innovatory concept for the watch industry. The world's top snooker champions, Dennis Taylor and Steve Davis, both sport Ebel watches, and so do Formula One driver Nikki Lauda, Three-Day Event Olympic Medalist Virginia Leng and world champion tennis player Stefan Edberg. Sponsorship has extended into the realm of culture with the involvement of such luminaries as conductor Leonard Bernstein.

As with the Beluga line, the straps, bracelets, clasps, dials and bezels are interchangeable, with two to four different sizes for men and women. The basic stainless steel Sports model is distinguished by its two-piece case, the movement held in by the bezel and five very small screws with the recessed setting crown balancing the screw set by the nine hour mark. The white dial has black roman numerals and optional date and seconds functions. The curve of the watch case is cunningly echoed in the interlinked 'wave' bracelet, an Ebel trademark, consisting of more than 190 different hand-assembled parts. Those with more expensive tastes can choose to have their watch face incrusted with a ring of tiny diamonds and the dial display similarly enhanced. 18 karat gold or a combination of steel and gold for both watch and strap, available also in leather, are additional options.

The special feature of both the Beluga and Sports lines is their water resistance, all models guaranteed to 30 meters (100 feet) except the Sports Discovery which is resistant to 200 meters (660 feet) when locked (unlocked this is reduced to 50 meters),

ABOVE *Ebel's gold 1911 watch with a heavy gold integrated bracelet.*

OPPOSITE *Ebel's gold and steel chronograph with three subsidiary dials, again featuring the familiar screws on its case-front.*

EBEL
Automatic

and their ability to cope with the hazards of an active life. Stringent quality tests are performed for an entire week, after assembly; they begin by testing the watch in a water-filled cason with the pressure raised to three atmospheres (100 feet), subsequently drying it for 10 minutes at 60 degrees centigrade. The constant movement a watch experiences is tested by the movement simulators, while resistance to corrosion and natural wear and tear is monitored in a similar laboratory environment. The watch is subjected not only to a salt-laden mist to check how much it can resist corrosion, it is also vibrated for two hours and then driven through a layer of sand and gravel 60 times to test its ability to withstand scratches and blows. The bracelet's flexibility is also put under pressure by mechanical twisting and vigorous pulling, to determine potential breaking points.

ABOVE *Ebel's 18 karat automatic Chronograph.*

ABOVE *The Ebel Chronograph with perpetual calendar.*

Such tight quality control is a hallmark of the company, which has its own separate workshops for the manufacture of movements, cases and bracelets and final assembly. The wristwatches are either run on an integrated circuit with a trimmer regulatory device or a high frequency chronograph movement with an equivalent error margin of one centimeter in 300 meters.

The Discovery was made with the diver in mind. Quartz versions have an end-of-battery indicator – the second hand moves in four-second time jumps. Each five-minute marker as well as the second hand is highlighted with a fluorescent point to ensure perfect visibility in total darkness. The one-way ratcheted rotational bezel, always color coordinated with the dial, is a precision instrument for pre-setting diving times.

A particularly interesting addition to both the Sports and Beluga lines that may well prove to become a collectors' item is the perpetual calendar chronograph. A stopwatch function that can record elapsed time from hours to a tenth of a second is combined with an accurate reading of the day of the week, the date, month and phases of the moon, without forgetting leap years, all backed up by a power reserve of 48 hours. An advance on this type of complex micromechanics is the recently launched Voyager.

Pierre-Alain Blum does not intend to extend the range, preferring to maintain its exclusivity. He is keen to emphasize that, though the business uses up-to-the-minute technology, there are still areas in which manual labor is unbeatable.

In 1986 the self-styled Architects of Time celebrated their 75th anniversary with the launching of the 1911, a more sophisticated version of the Sports watch with a three-piece case and domed sapphire crystal (a difficult technical achievement). The company, despite its international profile, is still based where it began in the Swiss watch manufacturing heartland of La Chaux-de-Fonds. A measure of how far Eugène Blum's small company has come in the intervening years was his grandson's purchase and renovation of the Le Corbusier-designed villa La Turque, inaugurated in 1987 as Ebel's public relations headquarters. It is an impressive monument to Pierre-Alain Blum's extraordinary success story.

RIGHT *Ebel's water-resistant Sport model; note the ultra-thin case.*

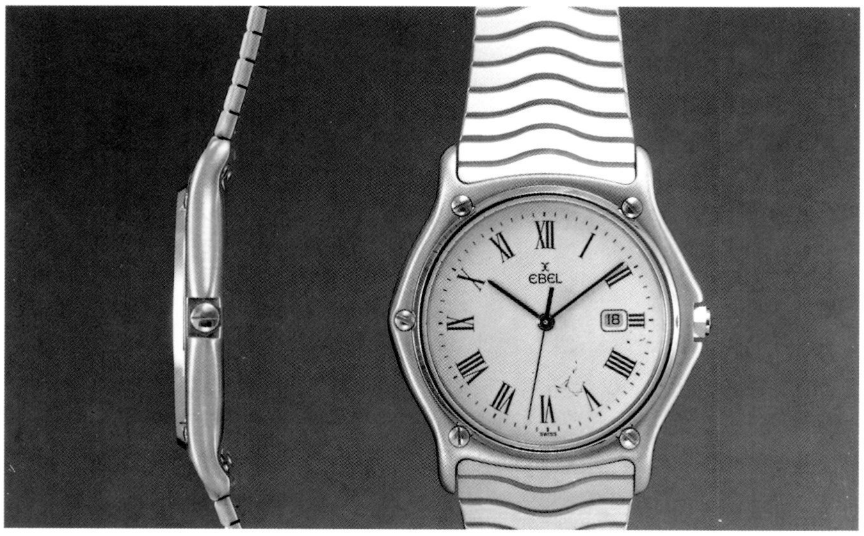

GEORG JENSEN

The name Georg Jensen has been synonymous with the best and most innovative in Scandinavian silver design since its inception in 1904. Just 20 years ago the company, now owned by Royal Copenhagen of Denmark, went one step further. Today Georg Jensen is still unique as being the only silversmith in the world to produce its own range of wristwatches, all of which have Swiss ETA quartz movements.

The original Georg Jensen (1866–1935) was an artist whose fascination and affinity with his craft resulted in the creation of outstandingly classic silverware. In the years prior to the First World War he had gained an international reputation as one of the important figures of the Arts and Crafts movement. His designs betray a remarkable sensitivity to form while the celebrated, almost matt-hammered finish became an international hallmark of the Jensen style.

'Silver', he wrote lovingly, 'has such a beautiful moonlight gleam, like the light of a Danish summer night. Silver can be dusky, and condensation can make it cloudy like a ground mist.' This remarkable man did not live to see the highly successful launch of the first Jensen wristwatch in 1968 in response to an increasing demand for watches that would complement the attractive elegance of Jensen silver jewelry. Designed by Vivianna Torun Bülow-Hübe, Sweden's first woman silversmith, the no. 326 watch rapidly became a byword in classic design. The organically curved open integral bracelet, manually shaped in stainless steel, terminates in a T-bar to facilitate removal from or replacement on the wrist. The crown is stylishly set at an angle for accessibility, while the plain mirror-like dial is a masterpiece of refined understatement. Later matt black versions (which like the steel variety are manufactured in four sizes, extra small, small, medium and large to suit every taste) strikingly sport gold hour and minute hands or glossy plain black hands set off by a gold central pivot. In 1975 Torun Bülow-Hübe, already a worldwide name with an impressive customer list that included Picasso, Ingrid Bergman, Duke Ellington and Brigitte Bardot, had the satisfaction of seeing this design bought for the permanent collection of the Bergen Museum, Norway. Nine years later the same watch was included in the design collection of the Museum of Modern Art, New York, a testimony to its classic modernity.

Georg Jensen has maintained its primacy in the world of silver by being open to new ideas. Since the death of the master, modern designers with widely differing artistic personalities and backgrounds have contributed their skills while still maintaining the traditional Jensen emphasis on outstanding workmanship and the highest quality.

LEFT *The designer of this Jensen timepiece is sculptor Henning Koppel. This austerely beautiful watch is available in steel or gold.*

ABOVE *The famous Georg Jensen bangle mystery wristwatch was designed by Vivianna Torun Bülow-Hübe. This is a future classic.*

OPPOSITE *Another entirely typical Georg Jensen watch.*

GEORG JENSEN

RIGHT *The unadorned elegance of Georg Jensen watches is exemplified here. The designers of this one are Torsten Thorup and Claus Bonderup.*

Although stainless steel is the predominant metal used in Jensen wristwatches, many parts of the watches are handmade, while the steel is treated and burnished with all the skill and care that is lavished on Jensen silverware. After all it has to complement top designer jewelry from Georg Jensen silversmiths.

Sculptor Henning Koppel was responsible for the design of another stainless steel classic in 1977, the no. 321 watch, which as well as being bought by the Danish State Foundation in 1980 has also found itself a niche in the design collection of the Museum of Modern Art in New York.

Koppel broke with the traditional method of numbered marking of time divisions (a numbered series in this model, marked up with roman numerals, is also marketed for those with more traditional tastes). Instead, the white dial is divided into minutes by delicately contrasting black minute dots and hands surrounded by an elegantly ridged steel bezel set off by a simple black leather strap. This beautiful design is also produced in stainless steel with a black face and contrasting hands and dial markings, in matt black with a white dial and black dots and hands and matt black with black face and white dial markings and hands.

A smaller version of each variety, the no. 320, also exists, which, owing to the proportions of dial to strap, is less impressive. As if aware of this, Koppel has launched a giant steel model with a diameter of 38mm – only for the big boys.

An 18 karat gold version of his masterpiece at four times the price was introduced in 1989. This boasts a sapphire glass on a tooled brown leather strap and comes in the traditional two sizes.

In 1985 the combined talents of architects Torsten Thorup and Claus Bonderup were responsible for the introduction of watch no. 347, another designer classic. Both the flat case and the woven strap are made of steel, polished to a semi-matt perfection with diamond dust. The rounded bezel contrasts with the square cut of the strap clasp and the watch lugs while a sense of solidity is balanced by the delicacy of the strapwork, the tiny dial markings and simple steel hands.

Only two years later the duo launched a further more exclusive version of the 347 in 14 karat gold on a lizard-like strap with matching gold dial and hands. A sporty version of the new 1347 in anthracite gray was also introduced, with a choice of a matching linked gray steel or leather strap.

A third architect, Jørgen Møller, is responsible for the latest Jensen model, which clearly owes much to the 321. The 351 includes two novelties, a date function and a sweep second hand. The date function on the left is balanced by the maker's name on the right, where it also remains for those versions that do not include this refinement. There are two sizes for each model, which is produced in steel and matt black with contrasting dials and seconds dots.

GERALD GENTA

Gērald Genta is one of the most exotic makers of wristwatches in the world today – and yet he is also one of the most respected for his craftsmanship and technical skill. This combination is rare indeed, and, because his output is only about 5,000 watches a year, they are highly collectable, as their occasional successful appearance in the saleroom demonstrates.

Genta's original designs illustrate perfectly the traditional link between the twin arts of the jeweler and the watchmaker, which remain right through to the retailing stage. It is no surprise to learn that he initially trained in Geneva (from the age of 15) for four years as a jeweler. After that he might well have remained out of view as one more person now competent in his craft. However, he then worked first in advertising and, almost unbelievably, in the *haute couture* business; this must have taught the young Gērald Genta something of the unusual demands of personal vanity, in the waspish and demanding world of fashion. Gradually his attention was drawn to the one luxury fashion accessory that moves all by itself.

During the following 20 years Gērald Genta designed wristwatches for most of the leading manufacturers described in this book; he was responsible for models such as the Bulgari, the Royal Oak (Audemars Piguet), the Nautilus (Patek Philippe) and the Titiane (Omega). By 1972 the time had come for him to strike out on his own account – to see his own name on dials of watches which he had designed. He bought factories in Geneva (for cases, dials and clasps) and in Le Brassus (for the movements), painted them pink, and started to assemble teams of outstanding craftsmen. A master Cabinotier Genevois was commencing work, and soon masterpieces began to appear. Some of them will never appeal to purist wristwatch buffs, but as brilliant fusions of those twin arts they have timeless appeal for many others.

Genta wristwatches may have quartz, mechanical or automatic movements, chosen precisely and with economics in mind: thus his perpetual calendars with moonphases offer choices of each, and come with varying dials and cases. A distinctive feature of some of his ranges is his use of the octagonal shape for the bezel, but with the line of each side very slightly curved, softened in a sense: it is (for reasons of good luck) particularly appealing to Asian markets, but also to sensitive eyes everywhere else. Purists should note that Genta has, in this shape, an automatic watch with a gold and steel bezel, a plain white dial, with baton numerals and hands and date indicator (at three o'clock). He has another automatic version in a round shape, made of gold, with black roman numerals, gold baton hands and sweep second hand, and date indicator at six o'clock; a different and gorgeous slightly mottled dial is also available. Collectors of skeletals should know that Gērald Genta has an elegant example (with black baton hands and crocodile strap) in his catalog; years ago he designed a famous skeletal wristwatch for Van Cleef & Arpels. In the heirloom department, Gērald Genta has a Grande Sonnerie pocket watch of great beauty and dazzling complication, and also minute repeaters; his children's department includes some expensive and delightful watches featuring Mickey Mouse and Minnie Mouse.

Gērald Genta is the only important watchmaker today who both manages his company and designs all its products. He has tremendous flexibility in his design concepts and established ranges; there is a definite high seriousness in the appearance of each model design which proclaims the distinct possibility (if there is no name on the dial) that it might be a Gērald Genta wristwatch. His Gold and Gold, Secret Time, Gefica Safari (with its compass on the clasp) and gentlemen's dress watches all possess that rare quality called style. And then there are his jeweled wristwatches. If the Duchess of Windsor was alive and in her collecting heyday, she would certainly be acquiring animal watches from the Menagerie collection: they are as captivating as any Cartier model. Genta's L'Esprit de Genève range of watches, each with a large semi-precious stone as a dial showing only gold hands and with jeweled bezels and bracelets, are certainly exotic; they are made in very small numbers indeed.

Collectors should note that all Gērald Genta wristwatches are numbered, starting at number one, with the exception of the minute repeaters, which have their own special series.

GIRARD–PERREGAUX

The place of this maker in horological history is secure, for it was in 1880 that Constant Girard designed and produced the very first production wristwatch. It was intended for officers in the German navy and its dial was protected with a four-by-four metal grille: not beautiful, but practical and, above all, new. If a genuine 'first' is a classic, then this watch has a welcome and special place in any collection.

Ultra-thin wristwatches have always been widely popular, and collectors know that it was actually the founder of Girard-Perregaux (in 1791), J.–F. Bautte, who produced in numbers the first ultra-thin or extra-flat watches. Innovative flair quickly became the firm's outstanding tradition. In the early 1850s the company that was to become Girard-Perregaux (in 1856) designed the extraordinary pocket watch now known as the Tourbillon With Three Golden Bridges; it incorporates the tourbillon invented by the master Swiss-born watchmaker Abraham-Louis Breguet (1747–1823), and won a gold medal at the 1855 World Fair in Paris, where Breguet himself had settled and worked. This outstanding pocket watch re-appeared in 1982 as an exact replica of the original, after years of planning the necessary 're-invention' of the movement.

In 1966 Girard-Perregaux produced the first high-frequency mechanical watch (36,000 vibrations per hour) and three years later the firm developed the first quartz watch to be mass-produced; the quartz oscillation selected (32,768 Hz) is now used as standard in all quartz watches. Contemporary Girard-Perregaux buyers will be familiar with the Equation range, which bids fair to enter the 'classic' category in years to come. It was created in 1985, and comes from a company which is the second oldest Swiss watch manufacturer, is privately owned, and has only 80 employees in its factory at La Chaux-de-Fonds in the Swiss Jura. Again technical innovation is the highlight behind the Equation range's appeal: the time standard and its divisions are electronic quartz. The longer periods of time are integrated and memorized by micro-mechanics: days, seasons, normal and leap years, moonphases, equinoxes and solstices, and periods of the signs of the zodiac (as in the Equation Espace Perpétuelle). All the watches' functions are easily controled by the crown, the battery operates for five years (whether or not the watch is worn), and, in a simple and imaginative design stroke, the back is clear glass so that the polished golden gear-trains on the circular brushed plates and the unique serial number of the custom-made ébauche are clearly visible.

LEFT *A fine gold modern automatic skeleton wristwatch from Gērald Genta, who is one of the most original designers at work today.*

LEFT *Modern wristwatches from Girard-Perregaux, with the date at twelve o'clock and peaked sapphires.*

TOURBILLON WITH THREE GOLD BRIDGES

Girard-Perregaux has for many years been following a policy of re-purchasing examples of its earliest classic watches. In the 1960s it bought back a famous watch – the Tourbillon with Three Gold Bridges (opposite). This gold hunter pocket watch has a white enamel dial with Roman numerals, Louis XV gold hands and a subsidiary seconds dial.

The company also decided to rebuild the 1880 tourbillon (invented by the master watchmaker Abraham-Louis Breguet between 1787 and 1795), but with 1980s technology, which was all that was available. The toolmakers and micromechanics had to work out the necessary, century-old techniques by dissecting the original movement, making detailed constructional drawings and recalculating every gear ratio.

The plates and bridges for the re-invented tourbillon were crafted in nickel silver because, like gold, the alloy does not oxidize and did not require electroplating. The color of the engine-turned parts harmonize perfectly with the gold bars holding the barrel, center wheel and tourbillon as well as the color-matched gold wheels. It is difficult to imagine the fine details of the mechanical work involved. For example, the detent arm has a rectangular section which, at its thinner end, is only 20 hundredths of a millimeter in height and ten hundredths of a millimeter in breadth. The detent has a ruby locking stone set in a hole four-tenths of a millimeter in diameter; the hole itself has a lining which is only seven hundredths of a millimeter thick. The chamfering of the parts and case is of a remarkably high quality.

André Curtit, a recent Curator of the International Museum of Horology in La Chaux-de-Fonds, has described this watch as 'the finest piece of work I've seen so far in my entire career'.

RIGHT *Girard-Perregaux's Equation Focale (left), with a peaked sapphire; the Equation Intégrale is on the right.*

RIGHT *Two perpetual multi-function chronographs from Girard-Perregaux's extensive catalog.*

RIGHT *One of the models in Girard-Perregaux's Equation range, the perpetual Espace.*

RIGHT *Two contemporary Girard-Perregaux wristwatches: on the left is the handsome automatic chronograph GP 7000; the dial comes in many colors, in a variety of case combinations.*

BASE 30 PULSATIONS

REVERENT
TRUSTWORTHY
LOYAL
HELPFUL
FRIENDLY
COURTEOUS
KIND
OBEDIENT
CHEERFUL
THRIFTY
BRAVE
CLEAN
BE PREPARED
A SCOUT IS

INGERSOLL

Those who remember the Ingersoll as the reliable low-cost quality British watch may be surprised to learn that both concept and company originated on the other side of the Atlantic. Despite somewhat unorthodox beginnings, Ingersoll's revolutionary success was achieved by turning a formerly crafted and hand-finished product into a mass-produced item, sold by an innovative and large-scale use of advertising. Both design and marketing approach were the brainchild of Robert Ingersoll, the son of a Michigan farmer, who had no previous experience of watchmaking.

The story starts in 1880, by which time the 21-year-old Robert and his younger brother Charles had moved to New York. In a short time they started up a general mail order company, specializing in a variety of goods uniformly priced at a dollar. Success was sufficiently rapid, no doubt as a result of Robert's entrepreneurial flair, that in 1892 the company felt confident enough to risk the launch of a pocket watch at a dollar and a half, the equivalent of a day's pay, at a time when a comparable model cost around $15. The Universal pocket watch, consisting of a tiny clock mechanism housed in a watch case complete with back winder designed by Robert Ingersoll, was a gamble that paid off with the aid of heavy advertising. As a result the Ingersolls decided to concentrate solely on the watch market.

LEFT *Two examples of Ingersoll's Boy Scout watches, made in the mid-1930s. These were some of the earliest 'own brand' labels used to merchandize watch products, the dials being printed paper glued on to flat metal disks.*

Three years later, Robert's ambition to create a dollar watch became reality and the Yankee was born. Later advertising puffs would talk of 'Ingersoll – the watch that made the dollar famous'. Year-long guarantees slipped into the back of each case were an early feature, establishing what was to become a hallmark of the Ingersoll reputation – reliability.

Their phenomenal success encouraged R. H. Ingersoll & Brother of New York to open up fresh markets and in January 1905, Robert and an old friend and business associate Estée S. Daniels sailed for Britain. It was not long before the 'Yankee' and the 'Crown' watch priced at five shillings had cut a swathe through the British market, spearheaded by intensive advertising targeted at both the public and the jeweler.

The retailer received point-of-sale material, sales training and a monthly magazine with up-to-the-minute information. By the 1920s Ingersoll's reputation for reliability was backed up by the addition of standard practice instructions and a school for training assemblers and repairers.

In the meanwhile, advertising copy relied on genuine testimonials from the general public as well as those in the public eye. Such intense advertising was unique for the period and a basic feature of marketing policy up until the 1970s. Ingersoll were also one of the first companies to take advantage of commercial television in the late 1950s.

The 1908 ladies' Midget was a prototype wristwatch. Smaller than the previous models, it was issued with a leather strap or holder. By 1913 lugs had been added to the watch case and the winding crown was moved from the 12 o'clock to the 3 o'clock position. This first wristwatch was manufactured in time to benefit from Robert Ingersoll's origination of the luminous radiolite dial, a feature that was found of inestimable value to the fighting services in the First World War, particularly in the trenches.

By the end of the war Ingersoll had expanded to own four factories in the USA and was trading on an international scale. However, by the mid-1920s the London branch was looking increasingly to Europe for supplies. The Wall Street crash of 1929 severed the already weakened link with the American parent, by then the Waterbury Clock Company. By 1930 Ingersoll Limited was launched as a British public company and sealed its independence with the opening of a large factory in Clerkenwell, London, equipped with advanced assembly processes.

In 1933 Ingersoll came up with another first, the character watch. Contrary to popular belief, the Mickey Mouse watch was produced before Ingersoll by Disney. It was followed up in 1960 by Dan Dare and Jeff Arnold watches, both characters featured in the popular boys' comic of the day, *The Eagle*. A readers' competition for a suitable advertisement for the watches was won by the young Gerald Scarfe, today a celebrated cartoonist. Runner-up with a commendation was none other than David Hockney, who has subsequently become one of Britain's best-known contemporary artists.

The advent of war in 1939 meant that the company's efforts were directed into the war effort. From watches for the services they diversified almost totally into instrument production which was to lead Ingersoll into high-security lock production.

The company's watch manufacture began again through an amalgamation with Smiths and Vickers at the suggestion of Sir Stafford Cripps, then Chancellor of the Exchequer, who saw the strategic necessity of a homebased watchmaking industry. In 1948 an Ingersoll factory was set up in Ruislip, north-west London, and the busiest period of the British company's history began.

By 1955, exactly half a century after Robert Ingersoll had arrived on British shores, production averaged a million watches a year, most being sold for under $9, though more luxury jeweled watches had been imported as early as 1925. A ladies' watch with seven jewels sold for just under $7 while a five-jeweled men's watch was marketed at just over half that price. Five years later the company could boast more than 100 different models, which included the enormously popular boys' and girls' watch range, waterproof and specially designed varieties.

After the Second World War, during which time a number of Ingersoll's own outlets were bombed, the sale of Ingersoll watches was restricted to jewelers only, which could but enhance the product's reputation for reliability and the jeweler's brand loyalty. But with the quartz revolution all this was to change.

The 1970s saw the influx of quartz watches onto the market. Ingersoll fell behind at this time, finding the competition uneconomical until the appearance of sophisticated microelectronics. The company, by this stage a small conglomerate with cutlery, electronics, printing, travel and merchandizing interests, was in its turn taken over by the Heron Corporation and lost much headway.

Fierce competition, lack of investment in advertising and the breakdown of the traditional relationship with the local jeweler network reduced Ingersoll's once impressive position. Today the company is handled by Steven Strauss & Co Ltd, who are rejuvenating the company, and it is fortunate that the assiduous and committed approach to marketing initiated by Robert Ingersoll and his successors has resulted in a legacy of goodwill and a reputation that endures to this day.

The Ingersoll Boy Scout pocket watch.

The automatic chronograph Da Vinci with perpetual calendar and moonphase indicator by IWC; this mechanical watch is programed until the year 2499. The retail cost in 1989 was nearly $14,000.

INTERNATIONAL WATCH COMPANY

The plain but ambitious name of this famous maker was dreamed up by an American, who most fortunately decided against using his own – Florentine Ariosto Jones! The new watchmaking business he set up in 1869 was located in Schaffhausen on the banks of the Rhine in German-speaking north-east Switzerland. Jones's inspiration for such a grandiloquent name is interesting to trace, and it led indirectly to a handsome annual income for a Dr Carl Gustav Jung.

F.A. Jones (1841–1916) had worked until 1867 for the Howard Watch and Clock Co., which had built its movement-making factory in 1857 in Roxbury (now part of Boston), Massachusetts. It was the first of its kind in America. At that time watchmaking in the USA was becoming a boom industry; the pioneer of mechanized watch production was Aaron Lufkin Dennison (1812–1895), who is sometimes called 'the Father of the American watch industry'. Dennison had moved to Switzerland in 1865, the first full year that peace returned to the USA after the trauma of the Civil War – paradoxically just before some of the most famous early US watch manufacturing companies were created. The American Watch Co., Waltham, Massachusetts (1859–1885), which became American Waltham Co. (1885–1921), was intimately associated with A.L. Dennison; it spawned associate and successor companies, such as Tremont Watch Co., Melrose, Massachusetts (1866–1868); The National Watch Co., Chicago, Illinois (1864–1874), which became the Elgin National Watch Co., Elgin, Illinois (1874–1954, and the name is still used today). The spirit of the age was one of innovation and expansion, with the three leading companies (Waltham, Howard and Elgin) together producing more than 100,000 watches in 1868.

In 1865 a simple event took place, which turned out to be crucial to the foundation of the International Watch Company. The pioneering A.L. Dennison moved to Zurich, to set up a branch of Melrose Watch Co., in order to take advantage of lower wage rates and local expertise. Dennison had previously traveled around Europe and was

OVERLEAF *International Watch Company's Flieger chronograph or Pilot's Watch, which was famous during the war and was relaunched in 1987. It features an anti-magnetic case of soft iron to repel electric currents that set up magnetic fields.*

IWC
SCHAFFHAUSEN

IWC
SCHAFFHAUSEN

confident enough to emigrate with his whole family, but, despite his great experience and contacts, the firm failed in 1868. In January 1869 F.A. Jones made his decision to move to Switzerland, to take advantage of his friendship with Dennison, and accept an offer of inexpensive premises in Schaffhausen from Johann Heinrich Moser (1805–1874), a watch and clock maker, whose hydro-power station on the fast-flowing Rhine and the cheap power it offered for orderly mechanized watch production greatly appealed to him. He took with him an old watchmaking friend, Charles Kidder, with whom he had worked for three years previously. At about the same time, Jones had noticed the foundation of the Illinois Springfield Watch Co., Springfield, Illinois, and he reckoned that his plan to export watches from the old world to these fledgling companies in the new world made business sense and fully justified the market-embracing name he invented for his new company.

Da Vinci – another version of this I.W.C. watch in steel.

Unfortunately, almost from the start, the new enterprise was not successful. IWC used brand names, such as Stuyvesant, on its watches, to find US markets, but Americans preferred own-country models; in 1864 the US government put a prohibitive 24 per cent import duty on complete watches; the initial investment in inexpensive Swiss labor began to fail, and Jones never made the 'break-even' number of complete watches, in spite of A.L. Dennison's presence in the background and all his advice. Within two years F.A. Jones urgently required fresh capital, and late in 1873 he set about promoting yet another joint stock company with new investors: the initial annual watch production was to be 10,000 units, enough to enable the company to make profits. A new factory building was commissioned, whilst problems with the supposedly inexpensive hydro-power to be supplied by J.H. Moser were being slowly sorted out. The economic scene had been shaken by the 1873 stock market and banks crashes in Vienna; America (his only planned market) was not taking the anticipated number of watches that Jones forecast in his financial prospectus, and the factory site and construction costs had been wildly underestimated. His co-directors gradually lost confidence in his managerial abilities, and in December 1875, just before the dawn of the age of the wristwatch, the International Watch Company was put into the hands of a receiver.

The special relationship with the watch industry in America was not, however, yet at an end. The company was bought from the receiver by a local banking consortium in order to save it from falling into foreign hands (still a feature of business life in Switzerland today), and this group promptly appointed another American to run it: Frederic Frank Seeland, who had worked with the American Watch Co., in Waltham, Massachusetts, and in London, re-established the factory in October 1876; but he spoke neither German nor French, and was incompetent. In August 1879 Seeland and his family suddenly vanished from Switzerland; an immediate investigation into the company's affairs discredited the modest profits of the two previous years and revealed dramatic stock and work-in-progress overvaluations. In November 1879 bankruptcy proceedings were opened for a second time. The American connection was finally ended; Florentine Ariosto Jones's brave dream was unfulfilled.

The second of the three chapters in the story of the International Watch Company is decisively headed 'The Rauschenbach Family', and once again J.H. Moser and the Rhine harnessed for his hydro-power play their central part. Moser had sold buildings and land in Schaffhausen on the Rhine in 1872 to Johannes Rauschenbach-Vogel (1815–1881), a successful engineer, engine manufacturer, industrialist and entrepreneur; the bankers had put him on the board of the International Watch Company after the first bankruptcy, and at the second he was left the main creditor. It was agreed that he should acquire the entire business to try and earn dividends for the other creditors, but a year later he died. His son and successor, Johannes Rauschenbach-Schenk (1856–1905), had the misfortune to go slowly blind during his short life, and he greatly relied on the abilities of Urs Haenggi, a thoroughly trained watchmaker and sound businessman who joined the company in 1883 and stayed with it for the rest of his life. He put the company firstly on to an even keel and on the road to successes which matched its name. One of them was the world's first quantity production of a digital pocket watch (Pallweber, 1884/5). Almost unbelievably, an American company once again knocked on the factory doors in Schaffhausen at the beginning

IWC
Schaffhausen
1986

RIGHT:
Porsche Design's renowned Compass Watch, which features an anti-magnetic automatic movement; made by International Watch Company.

RIGHT:
An uncomplicated mechanical with stylish dagger hour and minute hands and distinctive raised baton numerals; this watch demonstrates the diversity of watches from Schaffhausen.

of the 1890s: the Non-Magnetic Watch Company proposed a merger. Haenggi prevented his inexperienced and unqualified chairman from agreeing to this, and, as it turned out, the American company went bankrupt a few years later, creating financial problems for three major contemporary Swiss watch manufacturers, Aebi (in Bienne), Agassiz (St Imier) and Badollet (Geneva).

Electricity replaced pure water power in the factory in 1895, and production facilities were constantly being updated by Haenggi and a new, very talented, technician, Johann Vogel. By the turn of the century the 12.5 ligne calibers 63 and 64 existed, and the International Watch Company stood ready to supply the new market for wristwatches. During the First World War the company produced severely practical watches for the wrists of officers who needed synchronization and luminous dials.

It is at around this time that Jung's links with the concern began. The second daughter of Johannes Rauschenbach-Schenk, who died in 1905, had married Ernst Jakob Homberger (1869–1955), a Schaffhausen industrialist, two years previously, and in July 1905 he was awarded sole powers of attorney to act for members of the family; they naturally included the eldest daughter and her husband – Dr and Mrs C.G. Jung. Jung was practicing in Zurich as a psychiatrist in the years before the war, and was doubtless very glad of the augmentation to the family's income by way of dividends received. Indeed he wrote several times to Haenggi and Vogel saying so: 'Gentlemen', he wrote on February 8, 1911, 'Permit me on behalf of my wife and myself to thank you both for the encouraging results of the past financial year and for your competent and successful management. Yours respectively and obediently Dr C.G. Jung.'

E.J. Homberger's eldest son Hans Ernst Homberger (1908–1986) became, by inheritance, the last private owner of the International Watch Company in 1955; when the quartz revolution arrived in the early 1970s the company was already looking to different markets with new designs (some by Ferdinand Porsche), slimming overheads, advertising in export markets, and keeping in close touch with their bankers. The struggle proved too much, however, and in 1978 the company passed into the control of the West German VDO Adolf Schindling Ltd; a large conglomerate

International Watch Co
SCHAFFHAUSEN
SWISS

of watch manufacturers was to be created and floated off as a separate company, but finally only Jaeger-LeCoultre became a sister company.

This chronicle of the many financial vicissitudes of the International Watch Company serves to underline the fact that there has to be profit in the maintenance of traditional ways – 'yes, but will it sell' has to be the refrain. And now to some wristwatches. In 1884/85, the company began manufacturing, under licence from Joseph Pallweber of Salzburg, the first-ever series of pocket watches with digital time indications – hours on top, minutes below: a true first which later reached their wristwatches. The next classic came much later on, in 1940: it was the Fliegerchronograph or pilot's watch, with a large blackened dial, bold luminous sans-serif arabic numerals, hour and minute hands and a sweep second hand. The movement was protected from the influence of magnetic fields by an inner case of 'soft iron'; the extra long strap meant it could be strapped over a flying suit. In 1989 the International Watch Company launched the Aviator's Chronograph, which is said to feature the world's smallest chronograph movement for an analog display with a quick adjust device. This new version is stainless steel as before, has a 60 second indicator, with one quarter of a second accuracy, a minute indicator up to 30 minutes and an hour indicator up to 12 hours; its 233 parts are assembled by hand.

The Ingenieur range came in 1946 with a patented movement with two automatic constructions with click mechanism, limited rotor movement and an automatic winding mechanism. 1969 saw the introduction of the extremely collectable Da Vinci wristwatch, in an 18 karat yellow gold case, containing the company's first-ever quartz movement: its characteristics are the 'continuous' progress of the sweep second hand and the slight but audible 'whistling' sound of the tuning fork watches. The Compass watch of 1978 was the first to be designed by Ferdinand Porsche, of car fame, with moonphase, baton numerals and date display; the whole can be used as a prismatic compass. A year later the Titanchronograph appeared, again designed by Porsche; titanium was used for the case and bracelet for the first time. Today the flagship model is again an automatic Da Vinci chronograph (1986; about $14,000), with perpetual calendar and moonphase: unique features are that all display corrections can be made with the winding crown, and that it will run until the year 2499 (with adjustments to be made at a watchmaker's in the years 2100 and 2200).

International Watch Company's gentleman's wristwatch Titanchronograph, designed by F A Porsche with a titanium case and integrated bracelet.

JAEGER–LECOULTRE

An early 18 karat Jaeger-LeCoultre Reverso with a black dial and date aperture at six o'clock.

Stainless steel case, scratch resistant sapphire crystal, ultra-thin movement, integrated alarm – such phrases are scattered throughout the watchmakers' promotional literature, and trip lightly off the tongues of international sales representatives. So familiar are they that one takes for granted the practical advantages of these inventions, and forgets to inquire, who actually invented them?

Jaeger–LeCoultre, of course: one company that has never been satisfied with resting at the limits of the possible. The smallest mechanical movement in the world, the smallest quartz movement, the smallest analog chronograph movement – they all belong to Jaeger–LeCoultre. This company has an extraordinary record for research and invention that has benefited the horological industry as a whole and given a special resonance to the phrase 'Swiss made'.

Other companies have reason to be grateful for these inventions. Jaeger–LeCoultre is, quite literally, the driving force behind their own success. In the early 1900s, Jaeger–LeCoultre manufactured parts for Patek Philippe, Cartier and Omega, and today it supplies raw movements to leading names in the industry: Audemars Piguet, Chopard, Piaget, Vacheron Constantin, and its own sister company IWC.

When crowned heads, leading statesmen and other emissaries visit Switzerland, the gift presented by the Swiss government, as the pick of the nation's artistic and technological victories, is a Jaeger–LeCoultre Atmos Clock. Queen Elizabeth II, Winston Churchill, J.F. Kennedy, General de Gaulle, Haile Selassie, Ronald Reagan and Pope John Paul II all received one. This 'almost perpetual motion' clock is driven by thin air. Impossible? No such word in the Jaeger–LeCoultre vocabulary. Temperature changes of as little as 1°C cause a very volatile gas in an hermetically-sealed capsule to expand and contract, and the motion of the capsule is sufficient constantly to wind up the mainspring. The Atmos, with its very special low-friction movement, has a working span of at least 600 years, although atmospheric pollution necessitates a cleaning every 25-30 years.

At present, wristwatch movements require 100 times more energy than the Atmos to function – and, as environmentalists know, to manufacture even a standard battery requires 50 times more energy than the battery itself gives back. With the focus on energy saving and renewable sources, perhaps the Atmos technology will eventually be transferred to the wrist. Impossibilities have always been the raw material of Jaeger–LeCoultre's inventions.

They obsessed Antoine LeCoultre, who founded the firm in 1833. The date makes Jaeger–LeCoultre one of the oldest of the surviving Swiss watchmakers, and even today, when visitors arrive at Le Sentier in the Vallée de Joux and ask for 'the factory', they are automatically directed to Jaeger–LeCoultre. The large, present-day manufactory stands cheek-by-jowl with the small workshop which Antoine LeCoultre set up.

In 1833, his company was known simply as LeCoultre. It was more than 90 years before Antoine's grandson, David LeCoultre, joined forces with the Alsatian watchmaker Edmund Jaeger, and only after 1937 did all their watches bear the fine Swiss brandname. For nearly a century, the LeCoultre expertise remained at the service of other watchmakers in the Vallée de Joux.

Antoine LeCoultre invented his own machinery and tools to produce the top quality, high precision movements and parts that he supplied to other makers. Special milling machines were designed to cut wheels and pinions. But the invention which revolutionized the entire industry was LeCoultre's 'millionometer', the first instrument which was capable of measuring accurately to one thousandth of a millimeter. The benefits for precision manufacture were obvious, and this exceptional instrument caused the metric system to be adopted as the official measure in the Swiss watch industry.

Antoine won a gold medal for his inventions at the 1851 World Exhibition in London, and from 1847 to 1910 Patek Philippe selected LeCoultre components for his own watches. During these decades, pioneering technology resulted in the first watch with a crown instead of a key to wind the mechanism (1860), the first minute

JAEGER-LECOULTRE
REVERSO

ABOVE *A 1940s gold and steel Reverso, made by Jaeger-LeCoultre for Favre-Leuba, in a rectangular steel case containing the reversible gold-cased movement.*

LEFT *The famous Jaeger-LeCoultre Reverso, with its reversible case. Designed originally to protect the glass when worn by sportsmen, this watch has become world-famous since its resurgence of popularity in the late 1970s.*

and quarter repetition movements (1870), and in 1903 the world's flattest movement, a mere 1.38mm thick. This ground-breaking invention led to extra-flat chronographs and a super slim minute repeater wristwatch (3.2mm movement), launched in 1906. A self-perpetuating obsession with miniaturization has gripped the wristwatch industry ever since.

It was not until 1925 that LeCoultre merged with Edmund Jaeger. Technical facilities were then expanded, and the potential to manufacture a complete, autonomous 'Jaeger–LeCoultre' watch in-house existed for the first time. Inventions swiftly followed: in 1926 the first stainless steel case, and the duoplan movement (allowing a large balance for better time-keeping), and in 1929, scratch-resistant sapphire crystal and the 2 Ligne – a very special watch that had a success out of all proportion to its size. The 2 Ligne was, and still is, the smallest mechanical watch in the world. The 74 parts of its miniature movement are packed into a tiny space 3.4mm × 4.85mm × 14mm and, together with the dial, weigh less than one gram.

This diminutive masterpiece, which could be fixed as discreetly as a tiny clasp into a bracelet of pearls or diamonds, was well suited to the era of luxury liners and Hollywood glamor. Today, about 30 of these unique watches are produced every year, and the model has the royal seal of approval: Queen Elizabeth II owns one (white gold, with a diamond-encrusted case and band) and she wore it for her Coronation in 1953.

And what of the curious name? A 'ligne' is a unit of measurement (2.256mm) used to indicate the size of a movement. 2 ligne is thus the smallest ever made. Or as Jaeger–LeCoultre puts it: '. . . an object of exquisite taste, of great class, which exceeds the genius of its creator to such an extent . . . that he was unable to find a name for it'.

The fame of this watch has perhaps only been eclipsed by the Reverso, Jaeger–LeCoultre's rotating sports watch, which made its debut in 1931. The Reverso is now one of the company's bestsellers, and is one of several successfully relaunched classics. Both the 2 Ligne and the Reverso were certainly fashion novelties in their day, though the distinction between 'fashionable' and 'classic' hardly applies to Jaeger–LeCoultre models; the company has a knack of sidestepping transient or capricious design ideas.

The Reverso was tailored to suit the sportsman of the between-the-wars era, with a stainless steel case that pivots 180°, so that the dial and crystal can be turned face down; thus protected, it was a shockproof and corrosion-resistant watch, designed for the ski slope, the tennis court and the polo pitch. The distinctive rectangular dial and case tooled with parallel lines typify Art Deco elegance. When shut, the watch doubles as a piece of jewelry. In the 1930s, the case was often personalized with the owner's crest or coat of arms, and can still be engraved according to the customer's wishes.

During the Second World War the Reverso went out of production, but it was later rescued from obscurity by an Italian dealer, who discovered a handful of old, empty cases in a drawer at the Jaeger–LeCoultre factory. Fitted out with a new caliber, they immediately found purchasers back in Italy. The Swiss makers responded and in 1979 Jaeger–LeCoultre relaunched its own bestseller.

Today, Reversos account for some 2,500 of the 12,000 complete watches that leave the Jaeger–LeCoultre factory each year. This watch is entirely crafted in-house, and is considered to be the factory's most perfect example of technological and esthetic harmony, making it an essential item in any collection of classics. The new Reversos come in 18 karat gold or steel, with leather straps or bracelets and with or without diamonds. Colored dials, with leather straps stained to match, are also available. The dial is still defiantly Art Deco in style, but the quartz technology inside is modern (although a few mechanical models are also available). Since 1987, the Reverso has become water-resistant and there are now versions with a moonphase indication. A special two-tone steel and gold Integrated Reverso with an articulated gold bracelet was launched in 1983, to celebrate Jaeger–LeCoultre's 150th anniversary.

The 2 Ligne and Reverso did not exhaust Jaeger–LeCoultre's creative ingenuity, and world firsts kept coming. In 1953 the 'Futurematic' was born – the first fully automatic wristwatch, which dispensed altogether with a winding mechanism. If the watch stopped, a flick of the wrist was sufficient to get it going. In 1956 came the first

ABOVE *Another stylish Jaeger-LeCoultre Reverso.*

ABOVE *A gold-filled wristwatch by Jaeger-LeCoultre; it has a silvered dial with dot numerals and outer calendar ring, red-pointed sweep second hand and apertures for the day of the week and month.*

automatic wristwatch with an integrated alarm. Slimming records were broken with an automatic movement 2.35mm thick in 1967, and the world's smallest ever quartz movement, 1.8mm thick and 11.7mm across, in 1982.

Somehow Jaeger–LeCoultre never makes its earlier efforts redundant by new inventions. This is proven by the successful reintroduction of old models like the Memovox of 1951 and the rectangular Phases of the Moon watch of 1940. Both of these have recently been reproduced in limited editions, the moonphase to mark Jaeger–LeCoultre's 150th anniversary in 1983, and the Memovox to celebrate its own 35th birthday in 1986.

The Jaeger–LeCoultre archives carefully preserve the plans and ébauches needed to recreate these mature and complicated masterpieces, and collectors have to be swift off the mark. Although every single Jaeger–LeCoultre watch has the prestige of an individual number, there is a special cachet attached to limited editions.

Phases of the Moon retains the distinctive 18 karat gold rectangular case which, in 1940, was taking its cue from the shape of the popular Reverso. It has day and month indications and a subsidiary moonphase dial. 600 of these watches exist, each stamped with a laurel wreath and the date of the company's anniversary.

There are even fewer owners of the Memovox Jubilee. 350 pieces were made, honoring the Memovox's traditional style with two crowns, one to set the time and the other to set and wind the alarm. It is now the only wristwatch to combine an automatic movement with a mechanical alarm. In 1951, the Memovox was one of the very earliest alarm watches, and collectors may recall its signature tune: the distinctive, mechanical 20-second 'buzz' emitted when the crown was pushed in to wind the alarm. The Memovox Jubilee is water-resistant and comes in 18 karat gold or gold and steel with a champagne, gray or ivory-coated dial.

RIGHT *The Gaia lady's chronograph in 18 karat gold, featuring the world's smallest quartz chronograph movement and the Lepine pocket-watch style, by Jaeger-LeCoultre.*

RIGHT *Jaeger-LeCoultre's Odysseus collection features the Quantième Perpetuel chronograph; it contains over 250 parts and is programed without adjustment to the year 2100.*

JAEGER-LECOULTRE
AUTOMATIC

The success of Jaeger–LeCoultre's old models is matched by the new lines. After the Reverso, the current favorite is Albatross, dubbed the 'genuine factory product' since every single part of the watch is manufactured and assembled in-house. And that includes each of the 150 links, in 14 different sizes, which make up the unique and anatomically-contoured bracelet. Naturally this bracelet, introduced in 1986 for the Albatross II, is patented. The other design signature of the Albatross is its subtly shaped hexagonal dial which comes in three sizes, with or without diamonds. A 1983 Jubilee model to look out for is the titanium-clad Blue Albatross. This gun-metal blue and gold water-resistant watch would be at home, like its namesake, in the Pacific and southern oceans.

Classic simplicity in stainless steel from Jaeger-LeCoultre.

Two other modern classics should be mentioned: the Gaia, with a crown and hoop-shaped lug at 12 o'clock which transform the case into a 'stopwatch', and the Lyre, which is Jaeger–LeCoultre's no. 3 bestseller. The Lyre, so-called because the serpentine profile of the dial and integrated lugs resembles the curved horns of the archaic musical instrument, comes with baton or Roman numerals and with optional date and moonphase indications. The most expensive version is the ladies' model Lyre, with 83 diamonds on an 18 karat gold case.

For many, the classical simplicity of the Lyre epitomizes the Jaeger–LeCoultre style. But no company can afford to dismiss clear market trends, and at present these are toward increasingly complex technical watches with numerous indications, calibrated for sporting and executive lifestyles. 1987 saw the launch of Jaeger-LeCoultre's Odysseus Perpetual Calendar, incorporating an ultra-slim automatic movement programed until the year 2100 (no. 166.740.803). At a glance, wearers can read the time, date, day and month, as well as the year, decade and phase of the moon. Odysseus's memory is probably a little more sophisticated than its owners': the watch will never forget whether there are 28, 30 or 31 days in a month, or a Leap Year. By Monday, March 1, 2100, owners should have booked an appointment with a watchmaker to adjust the one-day discrepancy which will have accumulated as a result of the Gregorian Calendar.

In 1988, the Odysseus line was extended with chronographs incorporating another world first for Jaeger–LeCoultre: the smallest ever analog chronograph movement. This remarkable movement combines quartz technology with mechanical components, and took over three years to develop. Only 23.2mm across, and 3.7mm thick, it is one third of the size of a conventional mechanical chronograph and 40 per cent smaller than the smallest quartz chronograph. That has meant a new generation of elegant, slimline chronographs, including one for ladies a mere 30mm in diameter.

Odysseus chronographs have 18 karat yellow gold cases with distinctive pink gold 'ribs' clasping the bezel. There is also an alluring dusky version in tantalum – a material with the shimmer of black pearl. The functions are simple to operate with two push buttons, all models are water-resistant, and there is an in-built safeguard against unpopular trans-continental telephone calls: when you travel abroad, and particularly when you cross the international date line, you can set the chronograph dials to keep the time of the country you have left behind.

The new JLC 630 chronograph movement is now built into selected models of Jaeger–LeCoultre's technical triad – Odysseus, Kryos and Hera, which are all available in quartz, automatic/mechanical and chronograph versions. The Kryos is essentially a masculine chronograph, with a chubby, tire-like bezel notched from 0–60. This can be set to perform a countdown before races, for example, or to enable divers to check the minutes left before their ascent. Kryos comes in 18 karat gold, steel or two-tone, and with a gold, steel or untanned leather bracelet.

Hera, by contrast, has as much appeal for sporting women as for men – not least because Hera was the Greek goddess and protectress of women who spent her days plotting savage revenge on the many lovers of her faithless husband, Zeus. The top of the range is the JLC 630 chronograph – but all models share Hera's special feature: a gold bezel graduated to give you your pulse rate per minute (count 30 heartbeats, starting with the second hand at 12 o'clock). This could be a life-saver, particularly in a high impact aerobics class, where its wearer would not only be fit but both elegant and time-wise.

LONGINES

In 1927 Charles Lindbergh, the famous aviator, completed his conquest of the Atlantic in an historic flight lasting 33 hours and 39 minutes. During those long lonely hours, he dreamed of a timepiece for aviators, a wondrous device that would give the time in degrees of the arc and make it easier to read the longitude. What a boon for pilots such a watch would be. Lindbergh himself sketched the design which fired the enthusiasm of J.-P.V. Heinmüller, Longines' American director who, fortuitously, was also a pilot with a passion for aeronautics. Was it possible to make such a watch and could Longines, renowned as a timekeeper of sports events even then, do it? They could – and did. Lindbergh wrote personally to Heinmüller expressing his appreciation, declaring it would save 'seconds in obtaining a position'. From 1932 onwards, Longines' Lindbergh Hour Angle watch rendered great service to a whole generation of pilots in an age when modern navigational aids such as radio, radar and satellite were unheard of. But then, Longines have always made watches for special people and this is the basis of their worldwide reputation.

Longines' history goes back to 1832 when a young merchant, Auguste Agassiz, came to live in the watchmaking district of St Imier in Switzerland. He set up as an assembler of watches, farming out components to different workers, many of whom were craftsmen working from home. They finished the watches which were then sold through his company, Agassiz & Cie. This traditional method of watchmaking was profitable and worked well enough for a couple of decades. In 1854, however, the ailing Agassiz handed over the responsibility of his business to Ernest Francillon, his youthful nephew. Ernest quickly realized that quality could never be assured while all the products were finished by different craftsmen of varying skills, and were practically never identical. Obsessed by the idea of making all his watches under one roof and

Rodolphe is the name of the designer of this unusual dial from Longines.

taking complete responsibility for their quality, he built a factory at St Imier in a place called Les Longines (Long Meadows). Longines' early success owed much to the pioneering spirit of Francillon who was determined to push back the frontiers of what was technically possible.

Early on, Longines started making chronographs; the first came out of the factory in 1879. In 1912 the company met a new challenge by inventing the first automatic timekeeping device – the wire-cutting system, inaugurated in Basle at the Federal gymnastics meeting. Human error in timekeeping was eliminated as the athlete at the start of a race broke a tape supporting a weight which, in falling, caused a contact by cable with a chronograph at the finish. When the contestant crossed the finishing line, he cut a second tape which worked on the same principle. Bearing in mind that the reflexes of a good timekeeper, operating a chronograph manually, can vary between plus or minus 1/10th of a second, it is clear how this revolutionized the precision timing of sporting events. As a result Longines is now an official timekeeper of world sporting events such as the Olympic Games and Formula 1 motor racing, its name familiar to television viewers all over the world. What better recommendation for a timepiece can there be?

A sturdy yet elegant quartz Longines in their Lindbergh range.

Always in the forefront of new technology, Longines began making wristwatches as early as 1910. Its first men's wristwatch with a lid and small second hand achieved popularity in tragic circumstances – in the trenches of the First World War. Longines continued to make wristwatches of the utmost precision until the quartz revolution. Unpretentious and supremely functional, these early wristwatches have a timeless dignity and class.

Not surprisingly, considering the firm's inventive capabilities, Longines were responsible for the world's first quartz cybernetic watch in 1969. This was a considerable feat at a time when the module did not even have an integrated circuit; yet the quartz corrected the rate of a vibrating movement, achieving a degree of accuracy in timekeeping that had never before been reached by a watch of this size. Three years later in 1972 Longines launched its LCD (liquid crystal display) at the Basle Watch Fair, a quartz watch with crystal display developed with Ebauches SA and Texas Instruments which enables the hours, minutes, seconds and date to be permanently displayed.

Many of Longines' watch models are designed for the specialist. Its Conquest VHP (Very High Precision) is the world's most advanced wristwatch with an accuracy of about one minute in five years – five to ten times the accuracy of common quartz watches. This watch incorporates a new development, the thermo-compensated quartz movement, a clever combination of circuits that defeats the principal enemy of quartz watch accuracy – temperature changes. A distinctive timepiece in the sporty style popular in the late 1980s, it is characterized by 12 notches surrounding the dial, and a bracelet with alternating bands of different metals. It is available in various combinations such as titanium and gold, steel and gold, gold, and with a diamond-set bezel and bracelet. Moreover, it is also water-resistant. This is clearly a watch for the individual who values precision above all.

With the specialist in mind, Longines has developed this watch even further. The Conquest GMT, for example, has a rotating bezel and a 24-hour hand and can therefore display local time as well as that in another time zone. For the gambler, the Conquest Las Vegas is a must. On the reverse side of the watch is a mini gaming table. The Conquest 1000 Oe is designed for people whose profession brings them into frequent contact with powerful magnetic fields. Longines rightly reckoned that there were enough engineers, technicians and radiologists to constitute a market for this antimagnetic watch. A normal quartz watch easily withstands the magnetic fields generated by everyday objects like television sets or electric shavers, but technicians and scientists are often exposed to far greater magnetic fields. Shielding the VHP movement of the Conquest 1000 Oe against magnetic flux are two shells of pure iron which deter the magnetic fields. This watch was subjected to intense magnetic fields in the laboratory and only stopped when the magnetic field exceeded 1000 0e – although, when worn on the wrist, a watch rarely undergoes such exposure.

Longines' Conquest Quartz Chronograph is possibly the ultimate in sports watches. Obviously invaluable to the serious athlete or motor-racing driver, it can also be used

ABOVE:
A contemporary classic from Longines' Conquest line, with a titanium and gold case and bracelet.

RIGHT:
An early silver Aviator's Lindbergh watch from Longines, which is eminently collectable.

LONGINES

RIGHT *Longines' Charleston Collection features this straightforward gold-plated quartz watch with Roman numerals and a subsidiary seconds dial.*

RIGHT: *An early automatic Conquest from Longines, with mixed baton and Arabic numerals.*

ABOVE *A classic ultra-thin quartz wristwatch from Longines.*

by the jogger or anyone who wants to time an egg perfectly. As a watch pure and simple, it displays the hour, minute, second (small seconds hand) and date. Time zone adjustments can be made by moving the hour hand so there is no interference with the setting of the minute and seconds hands. As a chronograph, it records hour, minute, second and 1/100th second for periods of up to 12 hours; there are 30-minute and 12-hour counters. In addition up to 99 results can be stored in the watch's electronic memory of which the first ten are individually stored and can be called up one by one.

Longines' specialist watches exemplify the company's technical inventiveness but as well as these Longines produce a diverse range of precision wristwatches for the ordinary person who wants something 'individual'. Les Grandes Classiques are exactly what their name suggests – uncluttered, classic designs that will still look good in 50 years' time. The Charleston range includes replicas of Longines watches made in the 1920s and obviously reflects the popular nostalgic trend in watch styles. By contrast, the Rodolphe watches, launched in 1987, are 'designer' watches, aimed at affluent fashion-conscious young people. Named after its young stylist, this watch, in Longines' own words, is 'round, smooth and polished', and has been remarkably successful with a new generation of buyers worldwide.

Collection Mode deliberately follows changing fashion trends in haute couture and ready-to-wear clothes. These are youthful and stylish 'accessory' watches, complementary and rather understated in design. Perennial favorites among connoisseurs, however, are the Planetarium and the Complication. The Complication, which shows the phases and age of the moon, month, season, equinox, solstice and the signs of the zodiac, is indubitably a masterpiece, the ultimate of the traditional watchmaker's art. Its complicated movement, which is difficult and expensive to produce, was perfected at the turn of the century and this is one of the few watches with a mechanical movement that Longines makes. This watch perhaps sums up Longines' ethos, in which traditional craftsmanship and precision constantly strive to meet the challenges and needs of the 20th century.

ABOVE: *An unusual digital wristwatch from Longines.*

LONGINES
Automatic
Conquest
SWISS MADE

Mido
Commander
Mido
OCEAN STAR

MIDO

OPPOSITE *Two popular Mido watches, Commander (left) and Ocean Star (right); these come in two sizes and are water-resistant.*

ABOVE *A further Mido Ocean Star watch (for ladies) it has a steel case, is water-resistant, has quartz movements and is available in two-tone and gold-plated versions.*

The gentlemen's model Mido Ocean Star.

The name 'Mido' is practically synonymous with 'waterproof'. Since 1978 the company has focused exclusively on the production of water-resistant watches – a clever marketing policy which has enabled Mido to target the increasing number of water sportsmen, leisure-seekers and holidaymakers heading for long-haul destinations. The Mido wristwatch has found its niche as the perfect companion for windsurfing and scuba-diving executives. Nautical references on the watches themselves reflect this theme; a rope-motif adorns the bezel of the Cable watch, and an echo of Art Deco elegance can be found in the chevrons or stylized waves on the dial of popular Ocean Star models.

There is a history behind the outward-bound image of the Mido man and his timepiece. In 1934 the company launched a watch to brave the elements – the Mido Multifort; it became an instant bestseller. The first self-winding, water-resistant, shock-proof and non-magnetizable wristwatch, it was designed to withstand arctic or equatorial temperatures, tropical humidity, and every imaginable natural or man-made behavioral extreme. Anecdotes from satisfied customers helped create a Multifort mythology. This was the watch that an icecream seller in São Paulo plunged into icecream as a promotional stunt; that a gaucho strapped to the foot of his cow; that Monsieur le Comte S.B., a burly explorer, dragged through desert sandstorms and so on and so on.

The Multifort was also a favorite with Second World War heroes. It accompanied one American pilot on 68 air raids. When he was finally hit in the wrist by a bullet the glass protecting the dial shattered – but the movement kept ticking! Apocryphal or otherwise, the anecdotes testify, at least, to the international market that Mido commanded from its early years. By 1947, the Multifort was selling in 65 different countries, and in 1952 the Superautomatic model was promoted in advertisements designed by Salvador Dali (who had the right apocalyptic approach for the task).

Mido rightly rests on a reputation for design expertise, technical innovation, and above all, practicality and durability. When Georges Schaeren founded the firm in 1918, he made it company policy to carry out market research and establish the precise, practical requirements of clients before any watch went into production. His directors set out from the offices at Bienne, at the heart of the Swiss watch-making trade, and built up a worldwide network of personal contacts with wholesalers, retailers and clients.

Mido has continued to expand, and is still based in Bienne, though new factories were built in 1947 and 1963, and in 1972 the firm joined forces with ASUAG, Switzerland's large watch-making concern. This ensured Mido a position at the cutting edge of technological developments, and, together with the decision to promote water-resistant watches, has consolidated its position in the market place.

After more than 70 years of trading, Mido has moved on from the fashion-conscious timepieces of its earlier years to haute couture: a timeless, refined and understated look which characterizes the entire output. The easy legibility of the classic Mido face with elegant baton numerals, uncluttered by subsidiary dials and hands, is practically a formula – repeated with the subtlest of variations, on the Cable, Baroncelli and Commander watches, and other models in the leading Ocean Star series.

Early watches from the 1920s and 1930s, which perhaps show a greater variety of invention, are also keenly collected. During this era, Mido created baguette wristwatches encrusted with emeralds for the post-war female flapper, and pocket and pendant watches – some disguised as aeroplanes and footballs – reflecting the sporting life of the jazz age. The premier example is the Bugatti Mido, a wristwatch in the shape of a car's radiator which was created in 1930 for L'Associazione Automobilistica Knac. Early in January 1989, this piece fetched 44,500 DM at auction in Frankfurt. The Bugatti Mido was the first in a series of radiator wristwatches for motoring enthusiasts which are now classic collectors' items for the sportsman-horologist. The dial has chic Art Deco numerals and carries the Bugatti logo above 12 o'clock. Over the logo is the crown, cleverly placed to reinforce the radiator motif.

It was during the 1920s that Mido began the search for a watch to 'conquer the exigencies of modern life', resistant to shock, water, dust, magnetism, temperature fluctuation, perfume, oil and other chemicals. The Mido Multifort, produced in numerous models from 1934 on, was just the first of many breakthroughs which established the firm's technical reputation. In 1935, Mido launched an automatic wristwatch (not the first, but certainly one of the most reliable of early automatics). In 1936 came the Mido Permadur, a watch with an unbreakable mainspring, and in 1934 the invention of the Powerwind system for automatic watches, which coupled simpler construction with greater power reserve.

Mido's flagship invention, however, and the one which still dominates its image and advertising, was the Aquadura water-resistant system. Technical enthusiasts will know that the winder, or crown opening, is the achilles heel of the watertight watch-case. Mido stole a secret from the wine trade to solve this problem. Bottles of wine are laid horizontally to keep the cork moist. As it is organic material, this insures that it remains elastic and expands to plug the neck of the bottle, keeping the system airtight. Similarly, a specially shaped and lubricated piece of cork was used to plug the crown tube of the watch, protecting the movement inside from water penetration.

Mido's Ocean Star series, fitted with this revolutionary Aquadura system, was launched in 1959. Over the years, the popular collection has been constantly enlarged, and discerning wearers have noted a progressive refinement of detail. One watch to look out for is the Ocean Star No. 1, first launched in 1981 when Mido signed up Bjorn Borg, fresh from his Wimbledon triumphs, as its promotional ambassador.

The No. 1 models share the quartz ETA movement, Aquadura system, steel case and sapphire crystal glasses in common with most Mido watches. But the stylized 'waves' cut into the solid steel bracelet, set off against the simple rectangular dial, give a special fillip to the design (models 156 8712 for gentlemen and 153 7712 for ladies). The contrast of gold, navy and steel gray, and of matt and polished surfaces, is very striking. A novel feature, when first launched, was the seamless sapphire crystal covering the case. Detail on the plain dial is pared down to the bare essentials – not necessarily including even the baton numerals – making the No. 1 series the acme of elegant understatement, even by Mido's standards.

Another classic for collectors is the beautifully crafted Cable watch (1987), of which only 1000 were made. An anchor replaces the baton numeral at 12 o'clock, and a delicate rope motif decorates the bezel. For those who like the official seal of approval, the Mido Commander, another limited series, won first prize at the Swiss Watch Styling Trophy for automatic watches in 1984, underscoring Mido's combination of craftsmanship and esthetic finesse.

Mido is not totally exclusive, or high-priced, in its orientation, however. To service the young, fashion-conscious, cocktail-shaking sector of the market, Mido launched the Swing Line in 1986, creating a sophisticated and non-plastic answer to the ubiquitous Swatch. This series will not have rarity value, but Swing Line is already a minor classic of modern functionalism.

RIGHT AND OPPOSITE
Further examples of Mido Ocean Star watches.

Mido
QUARTZ
OCEAN STAR
SWISS MADE
7

MOVADO

Movado, a company whose name means 'always in motion' in Esperanto, began its life in 1881 in a small workshop at La Chaux-de-Fonds, Switzerland. Talented watchmaker Achille Ditesheim and his team of six employees made every watch by hand – a necessarily slow and expensive process. By 1890 Achille's two brothers, Leopold and Isidore, had joined him as partners and LAI Ditesheim, as it then was called, employed 30 people. The aims of the three brothers remain motivating factors in the company today; they wanted to concentrate on fine watches that would achieve an international reputation for excellence.

Recognition of their efforts followed, and in 1899 the company was awarded six first-class official rating certificates and 16 refining certificates by the Neuchâtel Cantonal Observatory. A prestigious silver medal from the Universal Exhibition in Paris followed in 1900. Encouraged and inspired by their early success LAI Ditesheim began to concentrate upon modernizing production methods. But technical expertise was not enough – the three brothers were equally concerned about good design.

They consulted artists, and set up a research department and drawing office as well as building a brand-new factory filled with the most up-to-date machinery they could find. Fresh capital was injected by yet another Ditesheim brother, Leopold's twin, Isaac, who was an engraver by profession.

The year 1905 represented a crucial turning-point for Movado, for it was then that the company chose this brand name, its 'hand' symbol, and won a gold medal and grand diploma of excellence at the Universal Exhibition in Liège. During the next nine years the firm went from strength to strength, and although this rapid progress was temporarily halted by the outbreak of the First World War, it afterwards continued apace. Indeed, Movado manufactured special military watches known as the Soldier's Watch, which were very popular at the time.

Movado was the first to manufacture 8½″ and 5½″ jeweled lever movements mechanically – these improved the accuracy of wristwatches, which were still a fairly controversial item. They won the Grand Prix at the Universal Exhibition in Brussels, and by 1910 were represented in Paris, Rome, Brussels and Rio de Janeiro.

The futuristic Polyplan wristwatch, launched in 1912, was another major achievement. Machines were used to make a movement with a profiled middle, which had mobile wheels and a second hand. Polyplan's revolutionary curved case fitted the contours of the wrist; these watches are now highly sought-after at auctions. Both Curviplan (1931) and Novoplan (1934) represent later refinements of this early design success.

Another Movado invention, the Ermeto (1926), is equally prized by collectors. The case is fitted with a device which allows the watch to wind itself automatically, as well as protecting the delicate mechanism from shocks, and temperature and pressure variations. These functions were certainly put to the test when Piccard and Cosyns took an Ermeto watch with them on their ascents into the stratosphere in the 1930s.

During this era American designer Nathan George Horwitt was exploring various ways to create an uncluttered watch face. These early experiments eventually led to the creation of the famous Museum Watch in 1947. His initial efforts resulted in what was probably the first digital watch, with a number for the hour and minute. But,

Horwitt says, '. . . the thing looked like a scoreboard.' He continues, 'It was in the right direction, definitely more direct and faster to read . . . though unfamiliar and perhaps less esthetic . . . There is a geometry (worth preserving) in the old-fashioned watch-face. It's been part of the experience since the sun-dial. That familiar space-time sequence is like recognizing high noon by seeing the sun directly overhead. We really do not know 'time' as a number sequence.'

The Museum Watch design is a stunning visual representation of these ideas. A single gold dot at twelve o'clock suggests the sun at its zenith, high noon. The moving hands, also gold, are set against a completely plain black face framed by a gold rim. Art and design experts were singularly impressed by Horwitt's outstanding achievement.

'I believe your design for the face of a watch is the only really original and beautiful design that I have ever seen. It places the element of design on an esthetic basic parallel to that of a painter's work,' wrote Edward Steichen, Director of the Department of Photography in New York's Museum of Modern Art. The famous American illustrator Norman Rockwell reacted in a more down-to-earth way, 'It is so damn'

LEFT *Movado's limited-edition collector's watch, the Andy Warhol Times/5.*

ABOVE *A gold day/date wristwatch by Movado; the silvered dial has dot and Arabic numerals, a subsidiary seconds dial and apertures for the day and month.*

MOVADO
MOVADO

MOVADO

original that I've never seen anything like it before. It is such a swell, modern, simplified design.'

Yet between 1956 and 1960 Horwitt unsuccessfully approached 15 watch companies with this design. In 1960 the Museum of Modern Art selected the watch face for its Design Collection, the first time a watch had ever been included as an example of modern design. By 1961, Movado had acquired the design – registering it under the name Museum Watch – and began to manufacture it. The present Museum Watch is, according to Horwitt, closer to his original design concept than the models displayed in the Museum of Modern Art and the Brooklyn Museum.

OPPOSITE TOP *Early mechanical Movado Museum watches.*

OPPOSITE BELOW *The current Movado Museum watch.*

'The model displayed . . . is not the ideal (case) design, but at the time no other was available. It has a curved crystal to give the impression of thinness that unfortunately made for highlights that went in direct opposition to the initial design concept which was to create a completely flat dial and a completely flat face (as in the present Movado Museum Watch).' The original mechanical movement has been replaced with quartz, which has considerably slimmed down the case size: an ageless classic.

A mid-1930s Tempomatic, an automatic watch with dagger hour and minute hands and a sweep second hand. The 'teardrop' intermediary baton numerals are unusual.

Movado's most recent venture into the world of art is equally progressive. This watch is unquestionably a collectors' item for it was made in a limited edition numbering only 250; 50 pieces are being retained by Movado themselves, while 200 went on sale for $18,500 each in the summer of 1988. This timepiece, the Andy Warhol Times/5, was the avant-garde artist's first watch and final work of art before his death in 1987.

How was this unusual project conceived? Andy Warhol was already fascinated by watches, and had a collection numbering around 300. As might be expected, his tastes were eclectic: '. . . he really got into watches in the mid-1970s. He was very knowledgeable . . . always looking, shopping . . . he was a very well-known face in auction houses,' said Vincent Fremont, the executive manager of Andy Warhol's Studio. 'He ran the whole gamut from junky, juvenile watches to very expensive. If he liked anything . . . he liked more than one of anything. Multiples.'

Warhol was also already connected to Movado through his long friendship with Gerry Grinberg, the chairman of North American Watch Corporation (Movado's parent company). In 1981 Grinberg suggested the idea of a limited-edition art watch, Warhol was enthusiastic – but nothing happened. However, as Fremont pointed out, 'Andy didn't do things by quickly going out . . . doing something just to put it out there. Sometimes he'd put things away for years . . .' When Warhol found a string of five watches wired together into a kind of bracelet, he began to see how he could create an exciting design of his own, and discussed the idea with Grinberg.

Warhol then experimented with the shape and decoration of the watchface, rejecting painted designs as unsuitable. He decided that the watch should have five working faces, and that he would use photographs to decorate them. Again, another period of selection and rejection followed.

About six weeks before he died, Warhol showed Grinberg some idiosyncratic shots of New York City and told him that these were the kind of pictures he wanted to use for the watch. His final picture selection and design for the Times/5 was found in an envelope after his death. Perhaps the last word should come from Theresa Morello, who had been making Warhol's photographic prints at the time of his death. 'For a man who loved watches . . . he thought about and worked on this for a long time . . . He wanted it to be special . . . five faces all working. I think he would be very pleased with what came out.'

OMEGA

Some companies have an enviable knack of being in the right place at the right time. At 10.56 pm Houston time, on July 20, 1969, it was Omega that first touched down on the moon. Neil Armstrong's Speedmaster Professional, the watch that timed his moonwalk, is still the official chronograph issued to every NASA astronaut.

It is Omega, too, that splits the seconds when Flo Jo smashes another 100 meter record. Since 1932, Omega has served as official timekeeper at every Olympic Games – an appropriate honor, perhaps, for a company with a Greek name that stands for the summum, or ultimate achievement.

Omega's Seamaster Professional is available in three sizes and is water-resistant to 200m (660ft). The one-way rotating bezel is notched 0–60 minutes.

Omega rarely misses an epoch-making event. Its association with grand achievements has attracted clients like Christian Lacroix, the Parisian fashion supremo. What hope, then, for lesser mortals? Recent years, in fact, have seen Omega work hard at placing its five leader lines in the upper middle, rather than the luxury, market for wristwatches. The majority of models in the Art, Symbol, Constellation, Seamaster and Speedmaster series now retail at between $500 and $1,500. Since 1987, this positioning has been backed up by the 'Significant Moments' advertising campaign, adroitly targeting the status conscious purchasing sector.

Any watch enthusiast will probably be familiar with these advertisements, which persuade us that moments of personal triumph are, if not of world-shaking importance, still worthy of an Omega wristwatch. The personal touch seems to have given Omega a clear edge over competitors in terms of brand awareness amongst the public. But any boasts of excellence are hardly idle. They come from a company with over 140 years of watchmaking experience, and an altogether exceptional record in precision chronometry.

Omega's present factory grew from an assembly workshop at La Chaux-de-Fonds, opened in 1848 by Louis Brandt. Here, Brandt manufactured key-wound precision pocket watches in silver cases which found a ready market throughout Europe. In 1877, he formed the Louis Brandt & Fils Company with his eldest son, Louis Paul, and in 1880 Louis Paul and his brother César moved the company to Bienne, where it remains today. There was a plentiful labor supply in Bienne, and the Brandt brothers began to manufacture all the components of their watches in-house. During the 1880s, the first brand names were launched, incorporating a novel cylinder escapement caliber developed in the factory. These names – Jura, Helvetia, Patria, Gurzelen – are an amusing reminder of a more parochial marketing era in the Swiss watch industry.

The company grew fast. Within a decade, it was employing over 600 workers and producing 100,000 watches a year – far more than many Swiss firms produce today. The Brandts reorganized manufacturing methods at their factory, introducing the 'divided assembly system' (like a conveyor belt system, based on the standardization of parts). Rival makers were quick to follow. The economies of the system meant that quality watches could be produced at a relatively modest price, opening up a vast potential market.

Using the new system, the Brandts launched their famous Omega 19 line caliber in 1894. In 1896, it was awarded a gold medal at the Swiss National Exhibition in Geneva. The name, Omega, was the inspiration of the Brandts' banker, Henri Rieckel, and it was chosen to emphasize the fact that the 19 line caliber represented the ultimate in watchmaking technology.

This watch, accurate and above all affordable, was a huge and a transforming success for the company. Louis Brandt & Frère (the company's name from 1891) became Louis Brandt & Frère – Omega Watch Co. in 1903, then Omega, Louis Brandt & Frère in 1947 and simply Omega Ltd in 1982.

This great 'O' (O-mega), last letter of the Greek alphabet and symbol of divinity, has been associated with infallible performance throughout the 20th century. The first Omega wristwatch, with a crown at 9 o'clock, appeared in 1902. In 1917 and 1918, Omega watches were chosen to equip combat units of the British Royal Flying Corps and the American army, and during the Second World War, the British government commissioned water-resistant steel wristwatches for crew members of the Royal Air Force – which led to commercial production of the Omega Seamaster after 1948.

OMEGA
Seamaster
PROFESSIONAL
200 M
OMEGA
Seamaster
PROFESSIONAL
200 M

OMEGA

It is hardly surprising that Omega wristwatches have been entrusted with coordinating decisive moments in man's history. Their records in observatory tests are unmatched. In 1963, at the Neuchâtel and Geneva observatories competition, Omega wristwatches beat every precision record in their category – the only time a single company has ever achieved this feat. At about the same time, NASA officials were casting around for a reliable wristwatch to coordinate the maneuvers of astronauts on the Gemini and Apollo space missions. A handful of top quality makes were bought anonymously from a Texas jewelry store. Only Omega's Speedmaster Professional survived the rigorous two-year program of tests.

LEFT *A gold/steel and diamond-set Omega Constellation Chronometer with calendar.*

LEFT *An 18 karat gold Omega lady's Symbol Classique.*

In 1965, the Speedmaster was confirmed as NASA's official chronograph. Five years later, the wisdom of the choice became clear, when explosions aboard Apollo XIII destroyed the spacecraft's timing instruments. With communications from NASA severed, the astronauts' wristwatches provided the sole – and crucial – link with earth time. To find the correct trajectory for a safe return to earth, the firing of the rocket engines had to be timed to a tenth of a second. The Speedmaster's lifesaving performance won Omega the 'Snoopy Award' – NASA's foremost honor. The watch also played a historic role in synchronizing the East/West rendezvous in space on July 17, 1975, during the Apollo-Soyuz mission: all the Russian and American astronauts, shaking hands on neutral territory, wore Speedmaster Professionals.

First created for sports use in 1956, the Speedmaster Professional is now universally known as the moon watch. For collectors, there is a special solid gold version, which combines the heavy-duty professional appearance of the dial with the allure of precious metal. The 18 karat case also has a transparent back, allowing a glimpse into history. To commemorate Neil Armstrong's touchdown, this version is engraved 'First watch worn on the moon – Apollo XI 1969'. As a symbol of peaceful progress, it should hold a significant place in any collector's cabinet.

The last two decades have seen a succession of technical and esthetic triumphs from the Omega factory. In 1969 came the Omega Dynamic, a distinctive elliptical wristwatch with a novel, streamlined dial and bracelet and hands of contrasting colors. The following year, at Basle, Omega exhibited prototypes of the revolutionary Megaquartz 2400. This watch, launched in 1974, was the first high frequency (2.4MHz) quartz wristwatch, and it has a precision ten times greater than any ordinary quartz model. The caliber 1511 version also holds a unique record as the only wristwatch to qualify in observatory tests as a marine chronometer. Megaquartz 2400 is a handy gadget for frequent fliers: the TSA (time zone and second adjustment) device will change the time zone without disturbing the minutes or seconds precision.

In 1980, Omega pulled its Magic watch out of a hat with the tap of a commercial wand. This super-slim bestseller has been a great success in the fashion market – and small wonder. Within the elegant rectangular case is set a completely transparent dial and inside the dial, two 'floating' gold hands keep time, with no apparent mechanism attached to drive them. For as long as you stare (and resist the temptation to have the watch prized open) it is impossible to see what keeps them moving. A sliver of a watch at 1.48mm thick, there is even a special collector's version of Magic, a mere 1.35mm thick. Those who prefer to double-check the time should probably opt for Omega's dual display Equinoxe instead. Unveiled in 1981, this was the first reversible wristwatch with

RIGHT *An Omega gentleman's steel- and gold-plated extra-flat De Ville watch.*

analog display on one side and LCD digital on the other as well as chronograph and alarm facilities.

Today, Omega's bestselling line is the Constellation, first introduced in 1952 and redesigned in 1982. There are numerous models, all of them identifiable by the current design signature: four 'claws', which grip the dial at 9 o'clock and 3 o'clock. Most of the men's models have subsidiary date and day dials and large Roman numerals engraved around the bezel. For ladies, a halo of diamonds around the dial (models C29 and C40) is an elegant alternative. The Constellations come in steel, 18 karat gold or steel and gold mixed, with an integrated bracelet or leather strap. All are water-resistant and selected models (C36, 37 and 38) have the transparent crystal back which is fast becoming a sine qua non for mechanical movements. Investors looking for something exclusive, however, can choose from top of the range models with natural stone dials in onyx, mother of pearl or lapis lazuli (C31, 32 and 33). Predictably, this series for successful executives has worked its way into the American business community, where it is worn as a badge of masculine achievement.

Also successful, but more overtly masculine, is the Seamaster series, targeted at divers and water sportsmen. This chunky, no-nonsense watch is made of titanium, a space-age metal twice as light as steel but equally tough and resistant to scratches and corrosion. The first titanium Seamaster was launched in 1982. It was water-resistant and had a screw-down crown. Recent models incorporate a thermocompensated quartz movement, subsidiary date and day indications, a luminous dial and hands, and glareproof sapphire crystal. The Seamaster Professional (models Sr 36–41) also has a 0–60 minute rotating bezel and is water-resistant to a greater depth. For those who perhaps identify with the image rather than the reality of diving, there is one tiny concession to fashion: the non-professional models also come with diamond chips instead of luminous hour markers.

The high-tech and overtly macho appeal of watches like the Speedmaster and Seamaster has perhaps obscured, or overshadowed, Omega's success in the ladies' and jewelry watch markets. In fact, Omega has won the most prestigious styling awards, through its association with top designers like Gilbert Albert, Luigi Vignando and Andrew Grima. Three Oscars at the Diamonds International Award in 1957, 1963 and 1964 secured Omega full membership of the New York International Diamond Academy – the leading authority on jewelry. The 1970s brought a string of awards including two Geneva City Prizes for LED digital display watches in 1975 and 1976, and several roses at the Golden Rose of Baden Baden for wristwatches by Luigi

RIGHT *A steel Omega automatic.*

RIGHT *An Omega Constellation calendar watch, with the days on the left and the date on the right.*

RIGHT *The Omega Speedmaster Professional. This is the watch that NASA selected in 1964 as its official wristwatch; Neil Armstrong wore a version of this hand-winding mechanical chronograph when he landed on the moon. The tachymetric scale is used for calculating speeds, rates etc.*

An early steel Omega Constellation Megaquartz with a date aperture at six o'clock.

Vignando. Vignando's Ramses II (1970), Salammbo (1971), Osiris (1977) and Structura (1978) models are classics for collectors of signature jewelry watches combining precious metals and stones.

Three technological breakthroughs also deserve a mention: in 1977/78 the smallest ladies' quartz movement in the world (caliber 1350) at that time, and the first 'baguette' quartz movement (caliber 1352), which redefined the shapes and designs achievable in the fashion watch market – as the baguette-cut diamond did for Art Deco jewelry in the 1920s; and in 1979, the Memomaster Quartz, the first ladies' multi-memory LCD wristwatch.

In 1987, Omega added the Art and the Symbol series to its leader lines, and both continue the company tradition of fine styling. The Art collection is a limited edition, with a choice of original designs on the reverse of the watch by Max Bill, Richard P. Lohse or Paul Talman. These miniature works of art are very much in the Bridget Riley 'Op Art' spirit, and the geometric style is a clever visual complement to the 'division of time' theme. Segments of color, arranged in rhythmical patterns, seem to pulsate and rotate – expressing time and motion and echoing the movement of the watch's hands. The colorful backs contrast with the stark black and white dial, which bears only the logo, a date indication and black baton hour and minute hands. Each piece is numbered and signed on the reverse, making this a series for collectors.

Finally, Omega has launched into the metaphysical realm with its Symbol series, based on the Sun and Yin/Yang themes. All these watches have a choice of multi-, two-tone or mono-tone dial, an 18 karat gold or two-tone water-resistant case, leather straps or bracelets, scratchproof sapphire crystal and quartz movements.

The distinguishing feature is the dial. On Yin/Yang models, a circle expressing the 'perfect whole' is divided into two equal halves, one dark (the Yin, which stands for the earth and female aspect), the other light (the Yang, which stands for the celestial and male aspect). Naturally the two are interdependent, so each half has a contrasting spot of the other as a reminder. In some models (S 18 and 19) the circle and bezel are incrusted with diamonds. The Sun dial is divided by a disk and radiating lines, which are intended to express the 'concrete' and the 'abstract' respectively.

Whether business, sport, or culture is your major motivation, Omega can now claim a leader line to match your aspirations. There are around 120 models in the Art, Symbol, Seamaster, Speedmaster and Constellation series, and they account for over two-thirds of Omega's sales worldwide. Altogether, some 250 Omega models remain from the 1000 or so available before 1985. This streamlining is a wise stratagy in an increasingly tough and competitive market.

PATEK PHILIPPE

What do Queen Victoria, Walt Disney and Albert Einstein have in common? They, along with many other rich and famous people, have all owned a Patek Philippe watch. This company, 150 years old in 1989, is one of the most prestigious watchmakers in the world. Both artistic and technical excellence have been its watchwords since the company began; Patek Philippe never deviates from the highest possible standards.

Antoni Norbert Patek de Prawdzic was born on 12 June 1812 in Piaski, a small town in Poland. As a young man he fought against Russia in 1831 in the Polish revolution. The following year, when Tsar Nicholas I crushed the revolt, he was one of the thousands who fled the country in fear of their lives. He moved to France, working as a typesetter in Cahors and Amiens. Settling in Geneva, he Frenchified his name to Antoine Norbert de Patek, and began studying art with a landscape painter, Alexandre Calame.

Patek became intrigued by watches at some point during this period, and put some together by buying first-class movements from master craftsmen. He commissioned goldsmiths, engravers, enamelers and miniature painters to decorate and make beautiful cases for these movements, and by the age of 27 he was a success. In 1839 Patek entered into two important partnerships: he married Marie Adélaïde Elisabeth Thomasine Dénizart, a French merchant's daughter, and formed a company with François Czapek – Patek, Czapek & Co.

François, who had also been involved with the Polish revolution, had studied watchmaking in Vienna and Prague before going to Geneva. Together with a small staff of five to seven they began to produce about 200 pocket watches per year. Some of these exquisitely crafted watches had repeat striking, and most were richly decorated. The embellishments show characteristically Polish themes: notably, the Madonna of Czestochowa and the Madonna of Ostrobrama are often found represented on the case backs. Each watch is numbered, the number punched into the bottom plate of the movement. On November 21, 1839 Patek and Czapek produced one of the first pocket watches which could be wound and set from the crown. Previously, a key had always been used to wind the mechanism.

But Czapek proved a difficult business partner, and Patek became restless. Then in 1844 he met Jean Adrien Philippe at an important trade exhibition in Paris. Philippe was a French watchmaker's son, and had been fascinated by his father's trade since childhood. After further training with a chronometer maker in Le Havre, and a period in London, he started a small factory in Versailles with the help of a loan from the French government. Of course, he had to repay this loan and needed to be profitable. He worked long and hard to perfect an extra-slim pocket watch with a

This highly distinctive Patek Philippe rectangular-cased watch features vivid gold baton numerals.

PATEK PHILIPPE
GENEVE

LEFT:
The five Patek Philippe watches shown (fourth from left is a Rolex) were auctioned at Sotheby's, London, on February 25, 1988.

ABOVE:
Four different shapes of Patek Philippe cases lined up for the same auction sale at Sotheby's, London.

crown winding mechanism. Rejected by a number of watchmakers, he took his innovation to the Paris Exhibition – where he won a gold medal and met his future business partner. Patek and Czapek had often disagreed with one another, and their association was amicably dissolved. The following year, on May 15, Patek & Co. was founded in Geneva, with Philippe as technical director of the new company.

The Victorian Industrial Revolution was affecting manufacturing of every kind. Watchmaking was no exception to this unstoppable trend, and Philippe set about modernizing production methods. In addition to inventing new machines, he patented his crown winding mechanism in 1845. During the next five years Patek & Co produced 2,618 watches – a notable achievement at a time when many craftsmen watchmakers were out of work.

By 1848 Patek decided to broaden his market, and began to travel widely, although not always willingly. In November 1854, for example, he wrote: 'My friends, the difficulties of the trip are beginning now. When will we be able to sell watches favorably and then wait for the customers at home, instead of having to travel all over the world with our products, incurring high costs and endangering our health?' Patek proved to be a reliable correspondent during these trips, and seemed to enjoy writing, for in 1863 he wrote a book entitled 'The Keyless Pocket Watches, that are wound and set without a key' and became a regular writer on the subject of watches for the *Journal de Genève*.

Patek Philippe & Co. was officially named in 1851, a move which recognized Philippe's considerable contribution to the firm's fortunes. The new name coincided with great success at the World's Fair in London, where Queen Victoria bought one of their pocket watches. This model hung from an 18 karat gold brooch decorated with 13 diamonds. Its ornate cover is enameled blue, with engraved flowers and diamond roses completing the embellishments on the cover of the gold case. It has ten jewels and, of course, Philippe's distinctive crown winding. Prince Albert chose a

ABOVE *A 1960s gold rectangular wristwatch with a textured chamfered bezel, signed Patek Philippe, Genève. The pre-sale estimate was $3,000–$4,500.*

ABOVE *A gold/steel rectangular wristwatch by Patek Philippe, the steel case with raised gold sides, the steel dial with baton numerals and subsidiary second hand. It nearly reached its top pre-sale estimated price of $2,250 at Christie's London sale on March 3, 1989.*

gold hunter with quarter-hour repeat striking, crown winding, and chronometer escapement. From now on Patek Philippe & Co. began to produce special watches for every exhibition and trade fair they attended.

We may think that cheap, 'pirate' copies of prestigious brands is a modern phenomenon – but as long ago as 1885 Patek Philippe discovered a forgery: a pocket watch signed Pateck & Cie., Genève. The extra 'c' in Patek gave the game away. A court case resulted, and it was decided that Patek Philippe should be paid the profits on these forgeries – 15,000 francs. The perpetrators of the fraud, Armand Schwob & Frère, were also forbidden to use Patek's name, however it was spelt.

The firm continued to expand. A new headquarters, six stories high, was built, and is still the headquarters today. In the leather-lined showrooms on the ground floor the firm's most expensive watches were displayed, along with all their medals and awards. By 1901, when Léon de Patek (Antoine Patek's son) left the company, the Patek family's involvement came to an end. However, two members of the Philippe family, Joseph-Emile and Joseph-Antoine Benassy, were still numbered amongst the seven directors who founded the new stock company, the Ancienne Manufacture d'Horlogerie Patek Philippe & Cie SA. Yet another storie was added to the headquarters in 1908, and an electric clock was set into the gable. This clock is now connected to the circuits of Patek Philippe's master quartz clock, the same model which also keeps time in the Vatican.

During the early years of this century as the wristwatch became increasingly popular, Patek Philippe began to concentrate their resources on developing mechanisms and designs for these then controversial items. A women's platinum wristwatch, with five-minute repeater, was made by the company in 1915, but only one was ever produced. In 1925 Patek Philippe launched a men's wristwatch with minute repeat. A range of 40 of these watches continued to be made until 1962.

By 1925 they had produced the first wristwatch with a perpetual calendar, basing the design on a ladies' pocket watch which had first been made in 1898. Production difficulties meant that this design had a short life, and it was not until 1941 that series production of wristwatches with perpetual calendars was economically feasible. This model, number 1518 (with chronograph), is now a collectors' piece, very popular whenever it comes up at auction. One of this series, made in rose-gold, was bought by the famous American boxer Sugar Ray Robinson in 1957. Sadly, it was stolen nine months later – a fact recorded in the discreet archives at Patek Philippe's shop.

The great financial crash of 1929 affected business all over the world. But out of this apparent disaster came one of Patek Philippe's lasting classics – the Calatrava (Model 96). This elegantly simple timepiece is still being made today – the only difference from the original being the frequently updated mechanism; the outer design remains the same as the 1932 original. The Calatrava cross became a recognized Patek Philippe symbol, and is found on a number of their wristwatches – although not all, as some people mistakenly believe.

How did this positive reversal of fate come about? The company, in common with countless others, found that there was simply no market for luxury goods during a time of severe economic recession. They were running out of money, and decided to sell off a majority shareholding in the firm. This was bought by Fabrique de Cadrans Stern Frères, which had long supplied Patek Philippe with raw movements. The company lost its last links with the original founders, for Jean Adrien Philippe – grandson of the first Jean Adrien – resigned. However, watchmaking is a business that seems to inspire family loyalty: two members of the Stern family are today president and general manager. They intend to keep Patek Philippe in the family.

Jean Pfister, the new chairman brought in by the Stern brothers in 1932, made an historic decision; Patek Philippe would manufacture their own raw movements instead of relying on outside suppliers. The Calatrava represents the 'new' company's resurrection, and the foundation of the company today. Pfister's second major innovation was to streamline production by limiting the variety of calibers to a select, first-class few. Running precision was thus vastly improved; Patek Philippe successfully entered around 500 watches in the Geneva Observatory's precision contests between 1944 and 1966.

OPPOSITE *A rare 18 karat gold minute repeating wristwatch made in 1961. The damascened nickel movement is adjusted for heat, cold and isochronism and has five positions, micrometer regulator, 29 jewels and silvered matt dial with applied baton numerals. At Sotheby's, Geneva, on November 11, 1986 it fetched Sw Fr 132,000 (including a buyer's premium of 10%).*

PATEK PHILIPPE
GENEVE

SAB
NOV

Technical innovations followed financial stability. In 1948 an electronics department was set up, which was to create the first fully electronic quartz watch without moving parts, and the first independent quartz pendule during the 1950s. Their first commercially-available quartz watch was launched in 1970. However, decisions taken during the 1960s are still adhered to today; Patek Philippe do not manufacture quartz watches with digital indication via light diodes or fluid crystals. Their quartz models with analog indication are offered as an alternative to mechanicals – both types are of an extremely high quality. They still produce about 20 to 30 per cent more mechanical watches than quartz models, most of which are made for the women's market.

Of the 40 Patek Philippe patents registered in Berne between 1949 and 1979, three are especially notable. In 1949, the Gyromax balance was developed. Not only did this balance improve precise regulation, but it could also be adjusted once it was built into the mechanism. In 1958 Patek Philippe devised a new method for attaching the hairspring to the balance block. The following year they launched a more efficient mechanism for changing local times when traveling, without moving the minute hand. Developments in the mechanical field have continued: the world's slimmest automatic movement (2.4mm), launched in 1977, is still in production. And in 1985 they added a perpetual calendar module to this movement, increasing its size to a still-slender 3.75mm.

Yet 20th-century technology, mass-production, and the twin evils of built-in obsolescence and disposability have failed to permeate Patek Philippe's exemplary standards. Production is limited to about 50 watches each day. These are lovingly created in every detail by the firm's team of master craftsmen; one watch may take from nine months to five years to produce.

Indeed, the attention to quality is unprecedented. Master watchmakers must undergo what amounts to a second apprenticeship once they join the company, while the chainsmiths will have trained for a total of eight years before being considered suitable employees. The company's enamelers are similarly among the very best representatives of their art. Any design or motif you can dream up can be painted on to the watch case by these talented artists – who often paint exquisite miniature pictures using just one hair. Patek Philippe are the only company in the world to offer this élite service to their customers. The engravers, goldsmiths, and jewelers are similarly gifted, displaying a formidable combination of artistry, training, and patient hard work. The result is that these watches are all subtly unique, for they are handcrafted right down to the last tiny screw, which is polished before taking its place in the mechanism.

ABOVE *A sturdy gentleman's gold Patek Philippe wristwatch.*

LEFT *These three fine Patek Philippe gold wristwatches are from the 1930s and 1940s. Note the stepped lugs on the left-hand model; at the Christie's, London, auction of March 3, 1989, it more than doubled its pre-sale estimate of $3,750. The case on the right is known as 'railway type'.*

OPPOSITE *Three very different watches by Patek Philippe, displayed for sale by Sotheby's, London, on February 25, 1988.*

RIGHT *An unusual lady's heavy gold cushion-shaped Patek Philippe wristwatch.*

Apart from the classic Calatrava, described above, there are a number of models particularly worth mentioning. The Golden Ellipse (1969) is available with quartz, automatic, or hand-wound movements. Its proportions were inspired by the Golden Mean – primarily an architectural measure of harmonious proportion which was employed in the building of ancient Greek temples and medieval cathedrals. The most popular design has a blue and gold dial; there are also diamond-set, white, and gold dials, and a skeleton model with a visible mechanical movement. Early Golden Ellipse watches had crowns decorated with the Calatrava cross, but this was later abandoned since it was felt to interfere with the design.

The Nautilus range have cases made from a block of metal – gold, or stainless steel. This means that they are water-resistant; to 60m/200ft for women's watches and 120m/400ft for men's. Most models show the date, and are protected by a sapphire

RIGHT *Two Patek Philippe wristwatches and a fob-watch on the right, which is an 18 karat gold perpetual calendar chronograph made 1945/50.*

A fine Patek Philippe gold wristwatch with perpetual calendar.

crystal. All mechanical Nautilus watches are made with automatic winding. Some Nautilus models have simple faces with baton numerals, and the date appearing on the right-hand side of the face. Others are lavishly decorated with diamonds and emeralds.

Recently, in 1985 and 1987 two multifunctional watches were launched. Model 3940 (1985) has a perpetual calendar, and indicates the moon's phase, leap-year, and the 24-hour clock. It is also the slimmest of its kind with a thickness of 3.75mm. The first series of Model 3970 (1987) sold out straight away – before the watches even reached the normal retailers. It has a chronograph, and perpetual calendar, plus leap-year and 24-hour clock indicators.

Another highly successful recent model is the 3919, with a porcelain colored dial, and Clous de Paris bezel. Its simplicity is dateless, and it has proved to be very popular. Fascinating skeleton models, where all the delicately complex workings may be seen, are also manufactured.

Confidential archive records are kept in fireproof safes at Patek Philippe's shop in the Rue du Rhône, Geneva. These books list all the company's famous clients – at least, all those who have purchased their watches at this luxurious address. Should one of these watches be stolen, this regrettable fact is also entered in the records. If the watch ever turns up again – perhaps for servicing – it can be returned to its original owner.

Of course, Patek Philippe also make very special watches – one of a kind – to order. An early example is a pocket watch they made for J. W. Packard, the American automobile millionaire, in 1927. Packard requested that his mother's favorite song should replace the usual alarm tone. The craftsmen succeeded, and for 8,300 Swiss francs Packard was able to hear La Berceuse from Godard's opera *Jocelyn* instead of a plebeian buzz.

On April 9, 1989 Habsburg Feldman, the Geneva auctioneers, held an exclusive Patek Philippe sale to celebrate the 150th anniversary of Patek Philippe; the last lot, number 301, was the Calibre 89, only just completed, and it fetched SFr 4.95m. This astonishing pocket watch is the most complicated the company has ever made – it has 33 complications divided into five main categories: timekeeping, calendar, chronograph, the chime and operational functions.

Patek Philippe watches are necessarily expensive to buy, since they are masterpieces of the highest quality. However, initial investment is rapidly rewarded. Watches with minute repetition have been known to increase in value as much as 1,000 per cent within a few years, while other models usually realize gains of around 100 per cent in the same timespan.

This elegant two time, gold Piaget has no main dial and two crowns. It is signed Piaget and Asprey.

PIAGET

As befits the director of one of the most exclusive watchmakers in the world, Yves Piaget spends more than 35 nights each year in a plane. A high-profile jet setter with a passion for equestrian sports, M. Piaget always launches a Piaget collection in style. St Tropez or Monte Carlo might be the venue and among the hand-picked guests who dance until dawn to the sound of champagne corks popping are international celebrities such as Boris Becker, Gina Lollobrigida, Gunter Sachs and Sammy Davis Jnr. This kind of high-level exposure is what the Piaget business is all about – selling watches to the rich and famous.

The rise and rise of the Piaget company, a relative latecomer in the watchmaking field, has a fairy-tale quality, appropriate enough since the business began on a small farm in the village of La Côte aux Fées (Hillside of the Fairies) in the Jura valley above Neuchâtel. In the mid-19th century Georges Piaget was one of a few farmers who eked out a meager living in summer and turned to tinkering with watches in the long lean winter months. In 1874 he established a tiny workshop where members of his family (he had 14 children) could work, at first part-time, turning out watch movements for different watchmakers. Over the next few decades the business became successful enough to occupy the family full-time. Only rarely in these early days did the company sell complete watches on the local market – marking Piaget and Co on the dials – and indeed the trading name was not even registered.

It was not until 1937 that the watchmaking business moved out of the farmhouse and into proper workshop premises, and only after the Second World War that Piaget established itself on the international market as a top-quality watchmaker. This was due to Gérald and Valentin, Georges Piaget's grandsons, who reorganized the company, created its first wristwatch collection and began marketing its products worldwide. Development from the 1950s on has been nothing short of spectacular.

In 1960 the family opened a jewelry workshop in Geneva (the movements were and still are made at La Côte aux Fées); in 1961 the firm opened its first foreign branch in Offenbach, West Germany, and in 1964 it acquired a majority shareholding in Baume & Mercier, the respected but ailing firm of Swiss watchmakers.

Piaget have always concentrated on *haute couture* jewelry watches and until the mid-1980s, ladies' models accounted for some 70 per cent of total production. Indeed, the company employs more jewelers than watchmakers, a fact explained by the richly embellished dress watches inset with hundreds of diamonds, rubies and sapphires that form a major part of Piaget's range. Add to this the matching necklaces, ear clips and rings available with certain models and one realizes that Piaget is as much in the jewelry business as it is in the watchmaking trade.

The company has led the field in producing dials set with precious and semi-precious stones. Not only diamonds and rubies decorate watches but coral, onyx, mother of pearl, opal and lapis lazuli – all distinctive, stylish materials – are also used. The Diamond Heart Watch, set with 132 stones, has a gem-studded dial, a heart-shaped bezel encrusted with 18 large diamonds and a matching diamond bracelet. This is of relatively conservative design, if such dazzling brilliance can be so called. By contrast, the sumptuous Galaxy, with its black onyx dial and sinuous gold and diamond case and bracelet, is streamlined and ultra-modern.

Piaget's forte as jewelry designers lies in the integration of watch and bracelet. Early on, Piaget realized that if its watches were to be worn as items of jewelry the bracelet must receive as much attention as the watch itself; in many models, such as the famous Polo, the face moves imperceptibly into the bracelet so that the immediate visual effect is of a heavy gold wristband which is also, almost as an afterthought, a timepiece. This is artifice at a high level, but so skillful is the design that it pays off. After all, as Yves Piaget nonchalantly remarks, 'You don't read the time from a Piaget – you admire it . . .'

Yet if this implies that Piaget are cavalier about watchmaking, nothing could be further from the truth. Like four other top names in the watchmaking world (Vacheron Constantin, Audemars Piguet, Patek Philippe and Rolex) Piaget have scored some notable technical triumphs. In 1959 it launched a ladies' watch with ultra-thin 9P

An 18 karat white and yellow gold quartz Piaget gentleman's wristwatch in a tonneau-shaped case with an integral bracelet, the horizontal bar design of which follows through the dial.

movement; in 1960 it created the world's thinnest self-winding watch movement, the Piaget caliber 12P with a thickness of 2.3mm.

Unlike many rivals who could not believe that the appeal of quartz precision would last, Piaget embraced the new technology enthusiastically. Indeed, in cooperation with an electronics research center (in which the company also has a shareholding) Piaget beat the Japanese at their own game to manufacture in 1976 the world's thinnest quartz watch movement, the 7P caliber. There is even a Piaget quartz movement, caliber 30P, with perpetual programing that will work until 2100. The only one in the world with the memory of the exact time. Even when the watch was switched off to save battery power, it automatically set itself to the right time when turned on again. Instantaneous time-zone change could also be achieved by using a simple switching device. Not only does it make automatic adjustments for leap years but it also takes into account the Gregorian calendar correction at the start of the next century.

Piaget's most enduring model is Polo, launched in 1980. A distinctive sports watch featuring horizontal gold bars on case and bracelet, it was an instant success and its avant-garde design much imitated. A water-resistant timepiece, it is now available in many combinations – interspersed with diamonds, with gold and stone dials, with round faces and diamond bezels or as a perpetual calendar watch. Similarly masculine and perhaps inspired by Polo's success is Chukka, with its heavy bracelet made up of gold wedges. The Diplomat, a classic design launched in 1965, is still a bestseller. Rectangular in shape with cut corners, the original has vertical fine *guilloche* lines continuing through the watch face. An elegant female version of this is now available with quartz movement, with gold and brown lines on the bezel and face.

Piaget continue to produce mechanical watches, which make up a third of the output, and the popularity of sports and 'daywear' models (including the new Dancer model) has meant a slightly lower production of jewelry watches. Piaget make only 12,000 to 14,000 watches per year, so when they do launch a new model, it is certainly stylish. One of the latest, L'Aura, is a jewelry watch and a symbol of Piaget's remarkable virtuosity. Made entirely of diamonds and coloured stones, it is as spectacular a watch as one is likely to see and an example of the company's sure touch in producing objets d'art which will become the rarities of tomorrow.

RAYMOND WEIL

Raymond Weil is the most recently founded independent maker of wristwatches to be mentioned in this 'A to Z'. After 25 years in the trade, he commenced his business, with Mme Simone Bedat, in 1976 in the village of Les Brenets, a few miles south-west of La Chaux-de-Fonds, and from the very beginning it was clear that Weil was going to take unusual routes to fame in the middle market for wristwatches.

Mozart and Vivaldi are his favorite composers, flying is his hobby, and his preferred writer is Antoine de Saint-Exupéry (1900–1944), the French novelist and aviator. It is therefore entirely to be expected that Raymond Weil travels very widely and that he has produced watches bearing names such as Fidelio, Othello, Traviata, Adagio and Amadeus. However, his chosen sector of the market is the quality fashion one, and he has attacked it vigorously, with clever marketing, advertising, sponsorship and special promotions. Older watchmakers are noting with interest, and perhaps envy, his progress, as his annual unit sales approach the half-million mark; retail prices range from around $187 to $1,700. He does not directly manufacture himself, but stays very close indeed to his suppliers (many of whom work almost exclusively for him); this is certainly not a new business concept, but it is generally a successful one, affording maximum flexibility whilst steadily approaching the stages at which some processes can be profitably brought in-house and under 100 per cent control.

RIGHT *Two shapes occur in Raymond Weil's Othello Collection.*

RIGHT *Raymond Weil's gold Fidelio wristwatch in gentleman's and lady's sizes.*

All Raymond Weil watches have quartz ETA modules (Flatline 3, and he has initiated his own 1.2mm module), with extra-thick 18 karat gold electroplating on the backs, bracelets and strap buckles; their glasses are either scratch-resistant mineral or sapphire, they are water-resistant, and come in ladies' and gentlemen's sizes. No unusual specifications here; it is the highly creative designs which make these watches stand out. For instance, the firm launched in 1988 the Traviata range – and, of course, there is a musical reference: the tuning fork. An integral strip of metal on the bezel at 6 o'clock starts across the dial towards 12 o'clock, but then separates into two parallel forks which finish at 11 o'clock and 1 o'clock. Specially colored and shaped mineral glasses are set into the divided spaces. On some models, all or part of the dial is pavé-set and the glass plain. On another model, a single integrated metal strip is set off-center from the top of the bezel to the bottom. The full pavé dials contain 443 stones, all set by hand. Raymond Weil's eight-style Othello range came the year before, in 1987, and, again, these modern, simple and ultra-thin timepieces have originality, instant appeal and lasting value (change the battery every two years, and check the bracelet or strap at the same time). Model 126P, without numerals and black-dialed, has a turban-like twisted black and gold bezel – very Othello.

RIGHT *Raymond Weil watches from their Traviata collection.*

Fidelio

RAYMOND WEIL
GENEVE
Othello

ROLEX

Readers of Britain's *Daily Mail* in 1927 were somewhat surprised by the front page news for Thursday November 24. The headlines were neither the latest information on the increasingly volatile situation in Germany nor concerning Trotsky and Zinoviev's recent expulsion from the Communist party. Instead the entire page was given over to an advertisement proclaiming 'the greatest Triumph in Watch-making'. A Rolex Oyster, 'The Wonder Watch that Defies the Elements', had seemingly performed the impossible. Worn by a young London stenographer, Mercedes Glietze, throughout her 15-hour 15-minute swim across the English Channel seven weeks previously, the watch had not only survived unscathed, it was still keeping perfect time.

Today, in an age when waterproof watches are taken for granted, it is difficult to appreciate the impact of such a story. But six decades ago this announcement was more than just a clever advertising ploy, it was the vindication of one man's dream.

Bavarian-born Hans Wilsdorf, orphaned at 12, was early taught the value of self-reliance. By the age of 24 he decided to set up his own watchmaking business, having gained experience in that profession both in La Chaux-de-Fonds and London. Wilsdorf & Davies was founded in 1905, registered in London, with a trade policy aimed at the specialty market.

From the beginning Wilsdorf had decided to concentrate in an area that was then viewed with a certain amount of derision – the wristwatch. In the early years of this century, when the larger pocket watch was still *de rigueur*, the wearing of a watch on the wrist was considered a mark of effeminacy. But even with the considerations of current taste and fashion put to one side, it was generally considered that the required size of movement would be insufficiently robust or accurate and easily damaged by its proximity to harmful elements such as dust and damp. Wilsdorf was unconvinced.

His first move was to lodge the largest order for ébauches recorded up to that time for a small lever escapement movement from Hermann Aegler of Bienne in Switzerland. Hundreds of models of the 'wristlet' watch, as it was called in those days, were tested in the Far East and British Empire markets, proving particularly popular in Australia and New Zealand. A series of silver ladies' and men's wristwatches were followed up by models in gold, all with leather straps. The expanding bracelet, now the hallmark of the Rolex watch, was added after 1906.

Watches during this period were still being issued under the name Wilsdorf & Davies, a company that in five short years had become one of the leading forces in the British watch trade. For generations the English watchmaker and importer had inscribed the product he was to sell with his own name. Wilsdorf had other ideas. He wanted to include a trade name of his own that would be short, easy to pronounce and remembered the world over. So the title 'Rolex' was born.

Acceptance of this radical break with tradition took many years. That is the reason why in the two decades prior to 1927 there are Rolex watches that bear both the name of the dealer and the new trade name, or the dealer's name alone.

That November 24 was a landmark in the history of the company not only for the launch of the first water-resistant watch – it also put the name Rolex irrevocably on the map.

BELOW *Five fine early Rolex Princes.*

Pages from an early Rolex Watch Co Ltd brochure promoting the Rolex Oyster, c. 1934.

ROLEX PRINCE
SWISS MADE
ROLEX PRINCE
CHRONOMETER
ROLEX PRINCE

ROLEX
PRINCE
SWISS MADE
ROLEX
ROLEX
PRINCE

PREVIOUS PAGE: *A further selection of eight gold or gold and steel Rolex Prince wristwatches. Note the elegant waisted cases on some of them. These watches are highly collectable if in good condition.*

ABOVE *Rolex's dramatic front-page advertisement for their Oyster wristwatch in the* Daily Mail *of Thursday November 24, 1927.*

ABOVE, RIGHT *One of the most famous of all collectable wristwatches, the Rolex Oyster.*

Protection of delicate watch movements had exercised the ingenuity of watch-makers from the early days of horology. Despite attempts to protect the movement with a dustproof band and Borgel's endeavor in 1891 to produce a single-piece case that screwed directly onto the movement, the possibility of water resistance still seemed Utopian. Indeed, for some, water-resistant models seemed an uncommercial proposition, fit only as 'web watches for ducks', as one wit remarked.

Wilsdorf's solution was threefold. A crystal specially adapted to the bezel precluded condensation. The casing, first patented in Switzerland in September 1926, featured a casing ring, back and bezel all threaded. When assembled all the case components were clamped together, thus rendering the joints totally proof against water and dust.

The third problem to be solved was to create an impermeable winding crown. Wilsdorf's Swiss patent for such an invention was issued only a month later than that for the waterproof casing. The mechanism consisted of the winding crown screwed onto an outer tube, which screwed in its turn into the caseband, and an inner sleeve threaded to receive the winding mechanism. The watch can be wound only when the crown is unscrewed, freeing the spring to engage the winding crown with the watch mechanism. When closed, the two metallic surfaces hermetically seal the watch case.

This basic system is still used in the classic Rolex Oyster watch of today. Guaranteed to a minimum depth of 100m/330ft, many a contemporary Rolex boasts an extremely hard and virtually scratch-proof synthetic sapphire crystal, while all possess the traditional tough casing, hewn out of solid ingots of stainless steel, platinum or 18 karat gold. The winding crown twinlock mechanism is assembled in 32 minute and precise stages. The triplock offers additional protection for the two deep-sea diving models, the Submariner (300m/1,000ft) and Sea-Dweller 4000 (1,220m/4,000ft).

Modern Rolexes are tested electronically in a dry medium even more stringent than the traditional testing in water. If a leakage is discovered, the Oyster case is submerged in an Étancheiscope, an instrument, like many of its specialized testing machines, developed by the company. A vacuum is created causing air bubbles to escape, thus indicating the precise location of the leak.

Wilsdorf was not prepared to stop there. In a prophetic letter written in 1912, he was to lay down the philosophy that characterizes the ethos of Rolex today. 'It is not with low prices', he wrote, 'but on the contrary it is with improved quality that we can not only hold the market but improve it.'

ABOVE *A stainless steel Rolex Oyster Perpetual Submariner.*

With the creation of the water-resistant wristwatch, Wilsdorf could turn his attention to the efficient solution of the self-winding watch mechanism. Originally invented in a crude form by Abraham-Louis Perrelet of Le Locle, Switzerland, in the mid-18th century, it was perfected in many details by the great French watchmaker Abraham-Louis Breguet, and issued under the name 'perpetuelle'.

Various watchmakers had made attempts in this direction, but with minimal success. Wilsdorf's contribution was the rotor mechanism, a metal mass of unstable equilibrium with the ability to rotate in two directions on a central axis which in its turn is connected to and capable of winding the movement at the slightest flick of the wrist. The mainspring is thus maintained fully wound to ensure lasting precision.

Rolex was also the first to include a date (Datejust) or day and date (Day–Date) function by windows cut in the dial. The latter, available in 26 languages, is a feature of the 18 karat gold and platinum models only.

For faultless accuracy the workmanship must be impeccable, a promise that comes with every Rolex watch. Every chronometer dial (out of a total of 41 models of masculine wristwatch in the Oyster range, 37 are chronometers) bears the legend 'Superlative Chronometer Officially Certified' and is accompanied by the Rolex red seal. This certification is testimony that the watch has been qualified as a chronometer according to the exacting standards of the *Contrôle Officiel Suisse des Chronomètres* (C.O.S.C.).

Rolex is no newcomer to such rigorous testing. In 1910, the very first Rolex wristwatch chronometer was awarded a first-class certificate. Four years later Rolex received the accolade of producing the first ever Class 'A' Timing certificate awarded

ABOVE *A Rolex Oyster Perpetual GMT-Master.*

BELOW *Four more collectable Rolex Oysters.*

ROLEX

OPPOSITE TOP *Two gold Rolex Prince wristwatches in waisted cases with protected crowns, separated by a square Rolex in a heavy yellow metal case.*

LEFT *Another selection of Rolex Prince wristwatches.*

by a testing observatory for a wristlet chronometer of 25mm (1lin) circumference, involving five positions and three temperature tests (ambient [around 18°C], freezing and oven-hot).

In 1925 Kew was again the testing observatory for an oval chronometer which also gained a class 'A' certificate, again the smallest contemporary wristwatch to receive such a result. Serial manufacture of chronometers began in earnest in 1936 after a special order of 500 Rolex chronometers for King George V's Silver Jubilee. The consignment was not only completed in 146 days; each watch also received a rating certificate with the comment 'specially good' from the Bienne station.

Such testimonials have kept Rolex in the forefront of horological science and Rolexes on the wrists of many of the 20th century's most famous explorers and achievers, such as landspeed world record-breaking Sir Malcolm Campbell, tennis star Chris Evert and voyager Tim Severin. A Rolex Oyster accompanied Sir Edmund Hillary to the summit of Everest in 1953 and Reinhold Messner's conquest of the same mountain without oxygen equipment 25 years later. Sir Ranulph Fiennes' TransGlobe Expedition tested the Rolex in polar conditions, while Commex, leaders in the deep-sea diving industry, automatically equip their divers with Oyster watches. The list is long and impressive.

Wilsdorf's early awareness of the value of advertising and promotion, so effectively used in the *Daily Mail* lead mentioned previously, has long kept Rolex in the position of a world market leader. More recently his successor, Andre J. Heiniger, who joined the company after the Second World War, has promoted Rolex in other fields. Golfer Arnold Palmer, conductor Lorin Maazel, opera star Kiri Te Kanawa and film director Franco Zeffirelli have all featured in recent Rolex advertisements.

It is probably less well-known that Rolex also manufacture a luxury line of watches – Cellini. Named after the celebrated Renaissance goldsmith, sculptor and autobiographer, Benvenuto Cellini, the 80-odd wristwatches in the collection are fashioned

ABOVE *Five circular-case Rolex watches. From the left: Rolex Oyster Perpetual Datejust, the same, Rolex Oyster Perpetual 'bubble-back', and two more perpetual date watches.*

ROLEX
PRINCE
CHRONOMETER

ROLEX
OYSTER PERPETUAL
CHRONOMETER

only from 18 karat white or yellow gold. Their production, no doubt, contributes to Rolex's high annual requirement of gold, making the business the biggest consumer of this precious metal in Switzerland.

Rolex do not advertise their connection with Tudor watches, a line introduced more than four decades ago to satisfy a market that required high quality combined with modest price. Nevertheless, their Oyster casing, together with day and date functions and rotor mechanism, leave no doubt as to their parentage.

Both Cellini and Tudor watches can come with a leather strap or bracelet. Not so the Oyster. Each is issued with a bracelet, the President for Day–Date models in matching 18 karat gold or platinum, the Jubilee for the Datejust in 18 karat gold, stainless steel or mixed metal and the classic Oyster, originally designed for the whole range and obtainable in a similar choice of metals.

OPPOSITE LEFT *A Rolex Prince.*

OPPOSITE RIGHT *A plain and simple Rolex Oyster Perpetual.*

Hans Wilsdorf died in 1960, when Rolex's name in the history of the wristwatch was assured. He had seen his empire grow from a small headquarters in London to an international concern, relocated in 1919 to the center of Geneva. The years since his death have seen further expansion, with the construction of a modern building in 1965 on the outskirts of the city that was effectively doubled in size by additions completed in 1981. Montres Rolex SA now own 19 subsidiary companies and have service centers in 24 of the world's major cities.

Continual striving for perfection has made Rolex watches sought-after collector's pieces. If Wilsdorf were alive today he would no doubt derive much satisfaction from the perusal of the watch sale catalogs from some of the major international auction houses where Rolex models of all periods vie with such illustrious names as Patek Philippe, Longines and Cartier. But would he really be surprised?

RIGHT *Roger Moore as James Bond in* Live and Let Die*; the Rolex played a vital part in the plot.*

ROLEX
DATEJUST
26

ROTARY

The Dreyfuss family have been making watches in La Chaux-de-Fonds for over 100 years, and today the business is still both owned and managed by them. Edward Dreyfuss, grandson of the founder Moise Dreyfuss, is proud of the family's continuing reputation for well-designed, good quality gold and gold-plated wristwatches at reasonable prices, and his own son Robert looks set to maintain and expand the ranges.

A London branch of the original company was founded soon after 1905 by Moise's oldest son Georges when he went to England; his brother Sylvain arrived a few years later. Together they stimulated world-wide demand for Rotary watches, causing a new factory to be built in La Chaux-de-Fonds in 1967. The third of Moise Dreyfuss's three sons, who had stayed in Switzerland, died in 1982 at the age of 93: his name was René. Soon afterwards the notable René Dreyfuss range of handmade mechanical watches was launched; this includes a gold-plated wristwatch, which is an hour and five minute chiming repeater (MP 006). They also produce two handsome gold-plated automatic skeletons (MP 008 and MP 009) which are less expensive. These watches are serially numbered.

LEFT *Rolex wristwatches, which feature continually in dealers' and auction sale catalogs.*

This Rotary advertisement photograph says much for the tradition of this old, established family company.

Tissot's PR-100, with step-motion sweep second hand and instant date-setting calendar (right).

TISSOT

For the past two hundred years, Switzerland's watch industry has been world-famous – synonymous with quality, precision, and style. Perhaps one of the best-known Swiss companies is Tissot, founded in 1853 by Charles-Félicien Tissot and his son, Charles-Emile, at Le Locle, where Tissot's modern factory is still located.

Charles-Emile had spent five years in New York; this had given him an innovative approach to marketing his family's watches, and the first beautifully crafted pocket watches were sold in the USA. Clearly a born traveler, Charles-Emile then set about selling Tissot watches to the wealthy imperial Russian market. His 52 visits were well rewarded, for until the Russian Revolution in 1917, Russia was Tissot's main outlet.

Charles-Emile's son, Charles, and his son, Paul, created their own revolution, which was to insure the survival of their company. Charles supervized the building of a new factory, and introduced machines that could do the work more efficiently than before. Paul refined production methods still further, and following family tradition broke into new foreign markets. By now Tissot was becoming well-known; some of their watches had won coveted prizes at the 1893 World Fair, Chicago; the Paris Exhibition, 1889; Geneva, 1896; and the Paris World Exhibition of 1900.

A sense of originality and innovation forms a continuous theme which links this company's early history with the present day. In the 1900s it launched a comprehensive range of wristwatches; in 1930 the Tissot anti-magnetic watch was the first of its kind. Tissot's Navigator watch was another first – a self-winding watch with universal calendar. (The latest quartz analog Navigator displays local time in all 24 time-zones.) The 1958 watch collection not only introduced the single caliber, but offered quality watches at realistic prices – an enduring policy to which Tissot still adheres.

The electronic quartz watch was a revolution in itself. Tissot's contributions to this field include in-depth research into the magnetic clutch, which facilitates accurate automatic time-zone changes, and the first analog quartz watch with combined multi-functional digital display.

Most recently, the Tissot PR Sonor contains a minuscule micro-processor which can change time-zones and program alarm times with absolute accuracy. The letters PR stand for 'proof': Tissot's PR100 range is water-resistant to a depth of 100m/300ft, and represents the top end of their traditional range of affordable classic watches in steel, gold plate and two-tone metal. They are still manufacturing charming pocket watches, too, although these days they are quartz rather than mechanical.

Tissot's latest innovatory achievement is the TwoTimer (1987). Newly perfected technology and highly-sensitive computer-controled machines were employed in its development and subsequent manufacture, reducing the number of parts by ten per cent and production costs by 30–35 per cent. The TwoTimer is a modern, multi-functional watch that represents another first in Tissot's pioneering history. Both movement and case are combined for the first time in a metal watch. This single piece is called the plate. Traditionally, the components of a watch are manufactured separately: the movement, case, dial, assembly and finishing are divided between different departments or even different companies. Charles and Paul Tissot would undoubtedly applaud the efficiency and economy of this recent invention.

The watch itself must be a modern collectors' classic for technological reasons alone. It is also symbolic of our data-hungry era, for its functions include the date, with relevant day of the week in English, French or Spanish; time in another programable time-zone; chronograph to the hundredth of a second; timer with visible countdown; and a 24-hour alarm. At present there are nine models, all water-resistant to 30m/100ft. As with all Tissot models, these watches are reasonably priced.

Technology aside, Tissot's most intriguing and unusual range of watches was launched in 1985. The RockWatch (a name coined by its inventor, Tissot's president, Dr. Ernst Thomke) is genuinely unique – for its case is made of Swiss granite, millions of years old, and no two pieces are the same.

The granite for these extraordinary watches is mined from quarries at Graubunden, Ticino, Valais, and Gothard-Furka-Julier. Different traces and striations of minerals and semi-precious stones are found in the rock, depending upon where it

A revolutionary technology is used for the manufacture of the Two-Timer, with both analog and digital displays; the frame of the movement is simply a plate, the edges of which merge into those of the case.

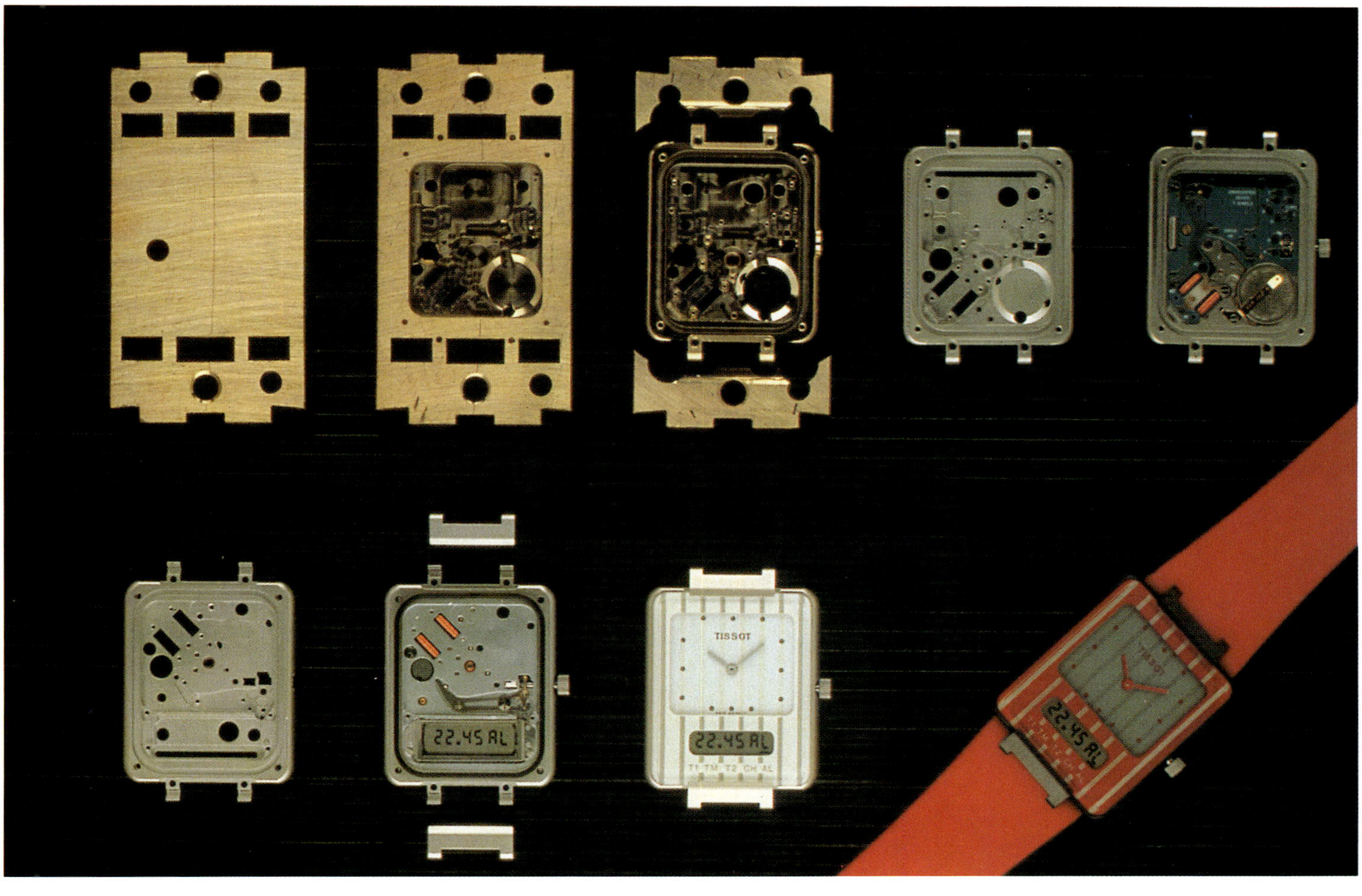

was mined. These include quartz, pyrites, topaz, garnet and tourmaline; tiny sparkling specks light up the gray granite watch face. Such geological variations also mean that the color of the granite ranges from dark, sombre gray to lighter tones with green or red overtones. Apparently, the vivid red and yellow hands were inspired by the stakes which mark out Alpine hiking trails.

While the TwoTimer reflects high-tech chic, the RockWatch is in tune with another worldwide modern obsession – nature. Since their launch they have clearly struck a chord, for more than one million have been sold to a public hungry for something different. This range of watches has been further developed to include 'Jewels of Nature' – watch faces made from semi-precious stones, basalt, mother of pearl, coral and shell rock.

In 1988 WoodWatch was launched – its satisfyingly simple case carved from ecologically-sound Corsican briar, a wood traditionally popular with pipe manufacturers. Certainly, Tissot have done much to make original design and reliable quality widely available. No doubt further originality and invention will carry them into the 21st century.

ABOVE *These Rock watches are a highly original concept and reach the short-list of many collectors.*

ULYSSE NARDIN

Ulysse Nardin is currently the astronomical and mechanical wizard among Swiss watchmakers. To the uninitiated, it is not a high-street name like Omega or Rolex. Nor does the firm service high-street clients. The Ulysse Nardin collector is a rarer species, with a mechanical heart and mind, and a deep pocket.

A prize Ulysse Nardin wristwatch, such as the Planetarium Copernicus (1988), retails at around 45,000 Swiss francs and the limited edition with six planets cut out of a meteorite would be considerably more. Those who covet the Planetarium, or Ulysse Nardin's Astrolabium Galileo Galilei (1985), do so in the spirit of the 18th-century collector who cherished the astrolabes, pocket globes, orreries and other scientific instruments in his cabinet as miniature mechanical embodiments of science, art, history and progress.

That said, Ulysse Nardin has not had a smooth commercial run in recent years. Its fortunes were nearly spiked by competition from quartz technology, and in the early 1980s, the company was on the brink of collapse. Salvaged in 1983 by a small group of investors under Rolf W. Schnyder's entrepreneurial direction, Ulysse Nardin made an overnight comeback with the presentation of the sensational Astrolabium in 1985. It is, once again, spearheading the renaissance of mechanical watchmaking. And now that we are on the crest of the 'Aquarian Age', and a renewed interest in astrology has rekindled nostalgia for moonphase and astronomical watches, the trading prospects of the company look secure.

Ulysse Nardin's headquarters are in Le Locle, high up in the Neuchâtel Jura. It was here, in 1774, that Ulysse's grandfather, Jean Léonard Nardin, set up a modest business making trustworthy stoves and water systems. His son Léonard Frédéric (born 1792) inherited these mechanical and manual skills, and became the family's first watchmaker, preparing the ground for Ulysse Nardin (1823–1876), the third generation. It was in 1846 that Ulysse set up the company, which swiftly won international recognition for the supreme precision and elegance of its pocket chronometers and alarm watches. Ulysse's watches were showered with awards, including the most coveted of all: the Prize Medal at the 1862 World's Fair.

Ulysse's exacting standards were respected by his son, Paul David Nardin, who took over the management in 1876. Paul David steered the business through its peak phase as the premier supplier of high-precision marine chronometers to international shipping lines. These chronometers, every captain's lifeline before the advent of quartz technology and navigational satellites, won Ulysse Nardin some 4,300 official awards from observatories.

After the Second World War, Ulysse's descendants began to specialize in slim automatic wristwatches, but the fast-moving, novelty-seeking market was slipping away from their grasp. When the marine chronometer was superseded by quartz and satellite technology, the generations of mechanical knowledge and skill invested in

The Ulysse Nardin Astrolabium Galileo Galilei.

Ulysse Nardin's Planetarium Copernicus. Production of this most remarkable wristwatch is limited to 65 pieces only.

the company's products seemed virtually redundant. The company's position was sorely shaken by the economic crises of the 1970s and the relentless advance of quartz technology. Attempts to seduce the Near Eastern market with wristwatches that offered plenty of gold but no 'unique' promotional features were an expensive failure. Stockpiles mounted, while the network of wholesalers and retailers fragmented.

In 1983, Rolf W. Schnyder rescued Ulysse Nardin from impending disaster. For 1½ million Swiss francs, he acquired a 60 per cent stake, and in recent years Schnyder has skillfully rebuilt the company's image.

With the unsaleable stock he inherited melted down, and a stable network of retailers found for some fine new mechanical wristwatches, the search was on for a totally unique, record-breaking wristwatch which would eclipse all competition and revolutionize the company's fortunes.

The Astrolabium Galileo Galilei, as aficionados know, did exactly that. In 1988, it made the front cover of the *Guinness Book of Records*. This unique astrolabe, a tribute to the inventor of the telescope, was publicly unveiled in 1985, at the Basle Watch Fair. It was the first wristwatch in the world to offer, in addition to the time of day, local time and month, many astronomical indications – at least for those who could understand them. Schnyder had to rewrite the operating instructions to make the many functions of this watch understandable for those who had no knowledge of astronomy. However, those who do not know one end of a telescope from another can easily appreciate the consummate craftsmanship of the Astrolabium, while for amateur astronomers it will indicate, at a glance, the ruling zodiac sign, the height and direction of the sun, moon and fixed stars, sunrise and sunset, dawn and dusk, moon phases, moonrise and moonset and the eclipse of the sun and the moon.

Different dials are produced to suit the owner's geographical location, and the watch is water-resistant to 30m/100ft. Around the bezel, the equinoctial and local hours are engraved in Roman and Arabic numerals, and the case is 18 karat gold.

Interesting patented features include the epicycloid mechanism with six simultaneous indications on a single arbor, and a special space-saving ball-bearing system to stabilize the mechanism. To guarantee shock protection, the movement parts are made of a light metal alloy, a third of the weight of the brass normally used.

This movement, a mere 9mm high, encapsulates the genius of Galileo Galilei. Once the cosmic science has been grasped, his learning does not weigh too heavily on the wrist. The Astrolabium was the brainchild of Rolf Schnyder and of Dr. Ludwig Oechslin, the archaeologist and scientific historian who developed the concept. And the story of its genesis is just as remarkable as the watch itself.

In 1978, Dr. Oechslin was commissioned by the Vatican to restore the Farnesian Clock, a complicated and magnificent 17th-century astronomical timepiece that had stood mute for more than 70 years. Since it had been donated to the Vatican, no one had ever been able to make it work. Unraveling its secrets took four years, and a comprehensive study of astronomy, mathematics, physics and philosophy. In the evenings, Dr. Oechslin earned his living as a cook at a restaurant in Rome.

Schnyder was first introduced to Dr. Oechslin's work at the studio of Jörg Spöring, where he spotted a 3-foot astrolabe which indicated the position of the stars as well as the time. With sound commercial instinct, Schnyder immediately saw the potential for an astronomical wristwatch. The partnership with Dr. Oechslin was forged – and remains active today (in his spare time, Dr. Oechslin, who is curator of the Swiss Museum of Transport in Lucerne, is shaping the future of Ulysse Nardin).

A prototype of the Astrolabium was ready in 18 months. First, however, its unwieldy thickness had to be reduced, to give it a chance in a market dominated by ultra-slim automatic wristwatches. Urs Gyger, the creator of the extremely flat 'Eterna-matic', was called in to revise the mechanism with Dr. Oechslin. The result is a miracle of precision engineering. The Astrolabium's mechanism is so accurate that, according to computer calculations, there will be only a one-day deviation from the exact position of the stars after a staggering 144,000 years!

Ulysse Nardin has not rested on its laurels, however. The company now produces around 40 mechanical models, and up to 3,500 pieces a year. For collectors and investors the choice is wide open, but since so few companies are capable of making them, the wristwatch minute repeaters must hold a very special place. The San Marco Automatons is unique among these, having a miniature reproduction of the automatons of San Marco's clocktower in Venice on the dial. Two tiny male figures flanking a bell hammer out the hours, quarters and minutes. Figures, bell, baton numerals and hands are all of 18 karat gold, exquisitely offset by the translucent blue enamel covering the dial. These figures and indices are supported on 32 micro-tubes inserted into the enamel, each with a diameter of only 16 microns. The dial alone requires more than 100 working hours.

In 1987 Ulysse Nardin created the one-minute tourbillon chronometer regulator with separate hour and minute indications, and the popular 'Michelangelo', a sophisticated 18 karat gold rectangular wristwatch with date, day, month and moonphase indications. Both these models have clear sapphire backs – showcases for the mechanical wizardry which is applauded by Ulysse Nardin's competitors as well as its clients. Like a piece of jewelry, the craftsmanship is always best judged by looking at the reverse.

For collectors, Ulysse Nardin's *pièce de résistance* is without doubt the Planetarium Copernicus. Unveiled at the planetarium in Lucerne in 1988, it is the first wristwatch in the world to display the entire solar system with the planets Mercury, Venus, Mars, Jupiter and Saturn. As the name indicates, the Planetarium is a tribute to Nicolaus Copernicus (1473–1543), the East Prussian scientist and canon whose revolutionary theory that the planets, including earth, moved around the sun, incurred the wrath of the Church. Copernicus's magnum opus, *De Revolutionibus Orbium Coelestium*, remained on the index of prohibited books until 1835.

The dial of the Planetarium combines Copernicus's heliocentric universe with Ptolemy's geocentric universe (based on the belief, which Copernicus rejected, that the sun, moon and planets revolved around the earth). The combination of these two systems, with both sun and earth motionless, makes it possible to track both the astronomical positions of the planets which revolve on separate rings around the sun, and also to calculate the angles between them, as measured from earth. It is these angles which give the astrological configurations, or the 'planetary aspects', which can be interpreted when charting a horoscope. The dial also indicates month, day, ruling zodiac sign and moonphase – and it is possible simply to read the time!

The mastermind behind the concept was, of course, Dr. Oechslin, and the Planetarium is his own tribute to the Swiss astronomer and inventor of logarithms, Jorst Burgi (1552–1632), who designed the first astronomical clock to combine the heliocentric and geocentric systems.

The rings of the Planetarium's dial are driven by a complex and ingenious mechanism developed by Dr. Oechslin with Bruno Erni, an ETA watch engineer. This movement, patented worldwide, is made up of 213 separate parts, and comprises

Another setting of Ulysse Nardin's Planetarium Copernicus.

a slim winding mechanism 3.6mm high and the Planetarium movement 3.25mm high. Together with dial and handsetting, the whole movement is only 8.5mm high, and, like the Astrolabium, the parts are made of an extra-light metal alloy to protect against shock. A unique friction clutch mechanism, preventing damage to the gearing system, provides an additional safeguard.

Prospective purchasers, and particularly mobile executives crossing the time-zones, may wonder how problematic it is to reset this gadget. But since all the planetary cycles are synchronized and geared into the single movement, the watch basically takes care of itself. If the watch has not been worn for a long time, it can be reset using the optional 'quick corrector' position of the crown. This means that the wearer need not wind the hour hand through infinite revolutions. Instead, the crown will control the revolutions of the moon around the earth – one turn being equal to a month. This advances the watch about 800 times faster than using the hour hand.

The seventh and outer ring of the dial is the perpetual calendar. One turn of this corresponds to a true year, or precisely 365 days, 5 hours, 48 minutes and 46 seconds.

It hardly needs to be added that the esthetic appearance of the Planetarium is equal to its technological wizardry. The case and the hand-engraved nameplates of the planets are in 18 karat gold, contrasting with the deep blue rotating rings, and the dial is protected by a domed sapphire glass.

A very special edition of the Planetarium, limited to 65 pieces, incorporates the material as well as the mechanics of outer space. The planet rings are cut from the meteorite found by Admiral Peary at Cape York, Greenland, in 1897. Most of these, however, are destined for museums.

VACHERON CONSTANTIN

How would you feel about driving around in a luxury Royce? Or shopping for all you need at Spencer, the highly successful retail chain store? Perfectly normal of course; but to read of what might have been is certainly a little puzzling! The fusing of the abilities and personalities of two gifted people can produce highly original and successful results which we identify totally with both names. After all, Audemars met Piguet, Patek met Philippe, Baume met Mercier, Girard met Perregaux, and so on.

Well-read watch collectors also know that Jacques-Barthélemy Vacheron went into partnership in 1819 with François Constantin (1788–1854), and indeed we know that he was relieved and happy with the new arrangements. He was thereafter able to devote his time to the problems and ambitions of the production side of the new business. His grandfather Jean-Marc (1731-1805), who founded his business in the St Gervais quarter of Geneva in 1755, and his father Abraham (died 1833), who succeeded Jean-Marc in 1785, both traveled widely, and often with the greatest difficulties in those stormy days, seeking orders and new markets. Constantin turned out to be an ideal partner; he was both an excellent businessman and an indefatigable traveler. Their new watches appeared in exclusive shop windows in America (where they had their own agency in 1864) as early as 1833, in Rio de Janeiro in 1840, and in 1850 they reached India.

In June 1839 a crucial appointment was made. Vacheron and Constantin took on Georges-Auguste Leschot as technical director. This man later came to be regarded as one of the founding fathers of modern watchmaking, because, with newly-developed part-making machinery, he was able to shake off the 'cottage industry' approach to assembling watches and bring a degree of standardization of ébauches and movements to hitherto time-consuming, and therefore expensive, manufacturing procedures. Then, as now, the company's policy was based on small production numbers and the highest quality. Movement thinness and classically elegant dials

An 18 karat two-color wristwatch by Vacheron & Constantin.

RIGHT:
TOP *A Vacheron & Constantin chronograph wristwatch; the silvered dial with outer telemetric and tachymetric scales, raised gilt Roman numerals for twelve and six, baton five-minute divisions, subsidiary dials for running seconds and elapsed minutes, and a sweep center second hand.*

RIGHT BELOW *An 18 karat gold Vacheron Constantin chronometer with subsidiary seconds and day and month apertures in red.*

MIDDLE *A gold square wristwatch by Vacheron & Constantin with a stepped case, brushed silvered dial and raised baton numerals and double baton quarter hour marks.*

LEFT *A rare gold rectangular wristwatch by Vacheron & Constantin with a raised curved glass; only 30 pieces were made in the early 1950s.*

were the outstanding features, and the company's traditional aims were well-known and secure as the dawn of the wristwatch era arrived.

In 1854, François Constantin died, and he was succeeded by his nephew Jean-François Constantin (born 1830); however, in 1867, in still unexplained circumstances, he reverted to being just an ordinary salesman, and César Vacheron succeeded him (only to die in 1869). Charles Vacheron (1845–1870) died after only a year in the chair, and was succeeded by his widow Laure Vacheron-Pernessin; she promptly brought in her 88-year-old friend Catherine-Etiennette Vacheron (widow of Jacques-Barthélemy Vacheron) and together they ran the firm for an astonishing five years. In 1875, Jean-François Vacheron was returned to the senior management, and in 1883 Catherine-Etiennette Vacheron died at the grand age of 101. The second old lady died in 1887, just as the first wristwatches were being privately worn. During this very unsettling period, the company was known variously as Abraham Vacheron-Girod, Vacheron-Chossat, César Vacheron & Co, Charles Vacheron & Cie and Vve César Vacheron & Cie. Finally, by 1896 both the directorship and the company's name stabilized: Vacheron Constantin it became once more, and is today.

The famous name first appeared on the dial of a wristwatch in 1910, joining the fashionable trend, and with it the Maltese Cross logo which had been adorning its pocket watches ever since 1880. This particular cross shape recalls a toothed wheel which was used in antique precision watches to regulate the tension of the spring. During the First World War, the company turned out military pocket watches and compasses, but thereafter the production of exquisite pocket watches continued, together with a modest output of wristwatches. In 1936, Charles Constantin assumed the presidency of the company, but renewed family pride was to be shortlived. The Second World War greatly reduced the company's fortunes, and Georges Ketterer took control in 1940. His son Jacques (died 1987) succeeded him, and in the post-war years, some fine and most collectable watches came from Vacheron Constantin.

RIGHT *Gentlemen's skeleton wristwatches from Vacheron Constantin in yellow gold or platinum, with three subsidiary dials and moonphase.*

Imagine a watch as thin as a toothpick! To mark the company's centenary in 1955, it produced a wristwatch with a movement just 1.64mm thick. Its 20.8mm diameter held a five-part bridge with over 60 parts to it, the whole being recognized at the time as the world's thinnest watch. Another potential entrant for the *Guinness Book of Records* came in the late 1970s, when Vacheron Constantin, shrewdly anticipating the worldwide publicity it would receive, produced what was at that time the most

LEFT *A gold Vacheron & Constantin lady's wristwatch with elaborate Arabic numerals.*

LEFT: *A gold Vacheron & Constantin wristwatch showing an unusual two-colored and textured dial.*

ABOVE *A white gold rectangular wristwatch by Vacheron & Constantin; the handsome white dial clarifies the baton numerals.*

expensive wristwatch in the world; it was called Kallista, which is Greek for 'most exquisite'. Its 140 gram case and band came from the same kilogram gold bar; they were then decorated with 118 emerald-cut diamonds, at the inspiration of Raymond Morette, a painter. He also fashioned the watch's unique logo and his signature above the hands; about 6,000 hours of work on this jewel/watch over about 20 months resulted in a sale price of $5,000,000.

Even today Vacheron Constantin produces only about 6,000 watches in its Geneva factory. Every watch drawing since 1840 has been kept, and there are full records of manufacture and sale dates, with movement and case numbers. Through many vicissitudes, the company has kept steadfastly to its founders' traditions; the finest of the older models feature in auction saleroom catalogs, and the newest and future classic wristwatches find ready sales through nearly 450 retail outlets around the world. At the top end of the price scale are the Lord Kalla and Lady Kalla; the second is apparently to be seen at special occasions on Princess Diana's wrist. It was a wedding gift from the United Arab Emirates; they were perhaps guided in their choice by Sheik Yamani, and it is indeed an outstanding example of a jewel/watch. It has 108 emerald-cut diamonds weighing a total of 30 karats on the dial and around the gold bracelet (the Lord Kalla has no less than 316 diamonds), and its movement is one of the smallest of its type in the world.

Vacheron Constantin's classic style for an everyday gentlemen's watch is seen in model 33060. This elegant mechanical retails at about $4,000; it has an elegant yellow gold case inclosing an ultra-thin 1.64mm movement; baton numbers, hour and minute hands, no second hand, no subsidiary dials, a white enamel dial decorated only with the gold Maltese Cross, no chronograph, no fussy lugs, and a plain black leather strap. This is the kind of wristwatch which makes a clear statement about its owner. There are two notable skeletons which should reach dealers' and saleroom catalogs in years to come, bearing in mind, as always, that Vacheron Constantin produce so few watches each year. Model 33014 costs around $9,350, and has Roman numerals hand-carved around the visible 14 karat gold movement; there are several models in this subtle range. The second skeleton is a grander, fuller affair. For less than $34,000, model 43032 brings you an automatic yellow gold or platinum watch with three subsidiary dials giving the day, date (including leap year), the month, perpetual calendar and moonphase.

ABOVE *An 18 karat gold wristwatch by Vacheron & Constantin, with a circular case.*

FAMILY FIRMS

ABOVE *Auguste Agassiz.*

With most old-established manufacturing industries, their history is social as well as industrial. This is particularly true of the Swiss watch-making industry, which from its very beginning has always been concentrated in small towns and villages such as La Chaux-de-Fonds, Le Locle, and St-Imier and in the mountains to the west of Bienne. In the early days tools and machinery were naturally primitive and the larger early makers relied heavily upon village inhabitants who were recruited as out-workers. Even today some of the major makers are surprisingly small, and subcontractors still play an important role in the manufacturing process.

One example of continuity is to be found at Longines. In 1782 Jonas Raiguel started trading in watches at St-Imier; in 1832 his son was joined by Auguste Agassiz, and in 1876 Ernest Francillon, Agassiz's nephew, built the company's first factory near St-Imier at 'Les Longines'.

ABOVE *Ernest Francillon*

RIGHT *Abraham-Louis Breguet (1747–1823), perhaps the greatest horological genius.*

ABOVE *Remo Bertolucci, one of today's leading makers.*

RIGHT *Longines' adjustment workshop at the turn of the century.*

OPPOSITE FAR LEFT: *Longines' first watch-making factory.*

OPPOSITE CENTER: *Longines' factory as it appears today, in one of Switzerland's small watch-making valleys, the Vallon de St-Imier.*

OPPOSITE LEFT: *Longines have long been associated with the timing of sports events; here the Italo/Brazilian Manuel de Teffe is winning the Brazilian Grand Prix in 1933 on the Gavea Circuit near Rio de Janeiro.*

The history of Swiss watch-making is studded with illustrious names which have survived on dials through remarkably long periods; Blancpain is one, and the company is described elsewhere in this book. Ownership of the company was handed down directly through nine generations. Also described in the 'A–Z of Important Makers' is the story of one of the most significant figures in watchmaking history, Abraham-Louis Breguet, many of whose inventions, such as the tourbillon, have remained unsurpassed to this day. Readers of this book will soon find that family histories are closely integrated with the development of the watch industry; many were the heirs who, just a century ago, were confronted with a whole new market – the wearers of watches on wrists.

Even today, new young master watchmakers are emerging, such as Remo Bertolucci and Gērald Genta. Both are heavily involved in their family-run businesses, exactly following the pattern of historical precedent. Like their predecessors, they take care to keep in the vanguard of developments in technology and design, and combine this with a strong awareness of market requirements.

ABOVE *The Omega factory in 1902.*

AMERICAN WRISTWATCHES

The watch industry at the turn of the century in America was enormous and was to spawn very fine makers. The establishment in 1882 of railroad time standards started the trend of popularity of watches for the wrist rather than the pocket. With this industrial growth the profit margins were never generally maintained; reinvestment in new machine technology and marketing techniques were limited. At the turn of the century the Swiss industry was becoming seriously established and their early 'bracelet watches', in elaborate and beautiful styles, successfully competed against the American mass-produced dollar watches. In the 1900s lack of tooling meant ladies' wristwatches predominated. The First World War boosted production of military men's watches, and government assistance enabled the larger firms to re-tool. The 1920s and 1930s saw the height of American wristwatch making, particularly in the Art Deco period. Thereafter, as the brief summary beneath indicates, only a handful of firms survived through to the 1970s.

BULOVA

Bulova was at one time the biggest maker in America. It is most famous for its Accutron (1960; production ended in 1976). Also notable are their 1920s and 1930s decorative Excellency range, and their highly collectable doctors' watch (circa 1930) and the Charles Lindbergh Lone Eagle watch (1927).

ELGIN

Elgin was founded in 1867 in Illinois. They introduced ladies' convertible wrist-watches in 1912. In about 1917 they produced a soldiers' watch with a pierced anodized case, outlined luminous Arabic numerals and hands with a khaki strap. In about 1926 came an attractive cushion form watch with an over-large crown. In 1928 and 1929 came their famous Art Deco ladies' watches, which are well worth looking out for; they include Madame Alpha, Madame Premet, Madame Agnès and Parisienne. Elgin produced its 50 millionth movement in 1951 but by then was in its last years.

GRUEN

Gruen was founded in 1874 in Columbus, Ohio, and had the advantage of Swiss family connections. More professional than some of the manufacturers, it marketed widely. In 1921 Gruen advertised a wristwatch as follows: 'Cushion, square strap watch, sterling silver, 17 jewel, adjusted, radium dial: $35 and $42.50. 14kt green solid gold: $65 and $75'. Heavily collectable are Gruen's famous range of doctors' watches brought out between 1929 and the mid-1940s; these had separate hour and minute dials with Roman and Arabic numerals, jump-hour variations available, in 14 karat white and yellow gold and gold filled cases. In about 1932 they launched the famous Curvex General, which had an extremely curved case to cover the top left half of the wrist. The Varsity, circa 1933, is also collectable.

HAMILTON

Hamilton was founded in 1892 in Lancaster, Pennsylvania. Its first wristwatch was produced in 1915, as a result of war requirements, and it has become perhaps the finest American wristwatch maker. In the late 1920s it produced enamel bezel watches: Coronado, Spur, Piping Rock. One of its most famous watches is the 1936 Seckron, a doctors' watch with separate hour and minute dials, in black and white, in an elongated stepped tank case. Two years later the Otis model was produced. In 1957 Hamilton launched its famous Electric Model 500, with a battery as a power source; then followed the asymmetric Pacer, Pulsar, Vega and Ventura. A strange but collectable watch to track down is the quaint, but very much of its period, Everest Electric (1958): it had a two-tone dial, but with the 12 on the bezel outside the dial, with the shape of the bezel area continued down onto the dial. The Thin-O-Matic came in 1963. Sadly the company was gradually broken up and sold in the 1970s.

ILLINOIS

Illinois was founded in 1869 in Springfield, Illinois. The company began making women's wristwatches in 1905 and men's in 1921. Illinois joined with Hamilton in 1927 and thereafter Illinois watches were Hamilton in all but name. It went on to give its name to some quality watches with offbeat designs: asymmetrical cases, very large curved models, Art Deco designs with auxiliary seconds in some models at 3 o'clock and 9 o'clock.

INGERSOLL

This maker is described fully in the 'A to Z of Important Makers' elsewhere in this book.

WALTHAM

Waltham was founded in 1885 in Waltham, Massachusetts; it first produced women's wristwatches in 1912, and men's two years later. Very collectable are their pierced case soldiers' watches (circa 1917–1919). An entertaining watch of about 1917 is the 20 Year Gold-Filled, which charmingly shows its pocket watch origins, as did so many of the early American wristwatches. In 1931 some interesting baguette models were introduced. In 1955 the wristwatch side of the business was sold off. Waltham figures in the story of the International Watch Company, which is described in detail in the 'A to Z of Important Makers'.

There are a number of other American wristwatch manufacturers with the occasional interesting product to look out for. Prominent among them are Agassiz (active 1915–1950s), Ball Watch Company (active 1950s–1969), Dueper Hampden (active 1921–1931 but then sold to Russia where it still manufactures), Hampden (1920s), E. Ingraham Company (comic character watches, 1930–1968), Mohawk (1930s and 1940s), Newhaven, New York Standard (mass producer of dollar watches), Rockford (1900–1915), Seth Thomas, Southbend (1903–1929), one of only two wristwatch manufacturers started after 1900 in America (which highlights the demise of the industry), and Waterbury Clock Company (out of which were born both Timex and Westclox).

BOTTOM:
A highly collectable Hamilton Electric.

COLLECTING WRISTWATCHES

WHEN YOU ARE PLANNING the formation of a wristwatch collection, there are obviously some basic rules to observe; most of them are simple enough and indeed can be applied to the collection of many objets d'art. Joining a club or society is perhaps the simplest way to begin (see Useful Addresses at the back of this book).

RIGHT *The diamond-set Omega Constellation.*

NEW WATCHES

These should only be acquired after careful professional advice from jewelers with a reputation they cannot afford to lose. In all capital cities there are a few great shopping streets lined with jewelers. However, if you are looking for a particular retailer's brand make, then you should shop in Dunhill for a Dunhill watch, Cartier for one of theirs, Hermès for Hermès, etc. Finally, remember to keep the box or packing in which the wristwatch comes to you: salerooms like these.

You might be considering a very expensive watch and it may be as well to double-check the retail price with one or two other shops before you purchase. When you have finally made your choice and buy the wristwatch, remember to keep the receipt (for later evidence of purchase and insurance purposes), and any certificate which may come with the watch. This might confirm the unique serial number of your watch; if, for instance, you have bought a Blancpain, you are invited to complete and send to their office in Le Brassus your entry form for the Golden Book, which the company has maintained since they started making watches in 1735. After you have done this your purchase will be recorded forever, and, moreover, you will always be welcome to visit their remarkable workshops in their village farmhouse.

ABOVE *Gentlemen's and ladies' sizes of two contemporary wristwatches from the famous Tiffany & Co, one of the great names among the world's jewelers.*

RIGHT *These fine wristwatches were auctioned by Sotheby's, London, on July 27, 1987. Note the rectangular waisted Patek Philippe with diamond-set numerals upon black.*

LUN
SATURDAY
28

FORTIS
SWISS MADE

Girard-Perregaux
automatic
Sea-Hawk

BUYING FROM SECONDHAND DEALERS

Unless you have become a considerable expert, it is important to know that the dealer has a good reputation and that he will not have vanished by the next morning. This means that you should beware of open market stalls, where ignorance, fakes and stolen watches lurk! If you find a watch that you really fancy do not hesitate to question the dealer closely about where it came from; attempt to check the asking price, possibly with a stall nearby, ask to see the case condition inside and out; if you are buying for the purpose of re-sale, note that a personal inscription can affect the value of a watch, unless it is dedicated to or from someone very famous (in which case it becomes an association piece). Ask the dealer to point out to you why the watch is not wearing out; ask him for any original certificate which might have the relevant serial number, and also its original box.

Finally, if it is a complicated watch you are buying, ask the dealer to demonstrate to you the working order of functions such as the chronograph, date, month and year changes and the progressing moonphases. Ask for a very full dated, signed receipt on the dealer's headed paper.

ABOVE LEFT *The dramatic Fortis Maxi-Classic watch, with a black dial and central moonphase; the date is highlighted in white in the aperture.*

LEFT *Two fine contemporary watches from Girard-Perregaux, the Sea-hawk (above) and the 7000 GBM automatic chronograph.*

LEFT AND BELOW *These wristwatches were auctioned by Sotheby's, London, on October 8, 1985, at a time when wristwatch collecting was becoming increasingly popular.*

CENTER *Five examples of the use of precious metals and jewels in wristwatch design; particularly striking is the Vacheron Constantin diamond pavé mystery watch (far right).*

AUCTION SALEROOMS

Until the 1980s wristwatches were not part of the clock and watch sales but were occasionally sold with jewelry. At the end of this book you will find the names and addresses of the leading auctioneers which hold regular sales.

It is important to go to the preview of the lots in the auction you are attending. Attendants will allow you to handle watches, under their supervision, which have caught your interest; read the catalog descriptions thoroughly, and, later on, thoroughly double-check the auctioneers' estimates of prices likely to be obtained: they normally give approximate minimum and maximum prices expected, but remember that they will generally be under-estimating rather than over-estimating the prices. At the preview, if it is your first auction, try and spend some time noting the particular interests of other prospective buyers around you. Do not be misled by the fact that most people will look at a watch which is attractively described in the catalog, with a well-known name and a very low estimated price.

On the day of the auction arrive early and listen around. By now you will have thoroughly read the catalog and the Conditions of Sale at the front of it; always remember that price estimates can occasionally be very wrong either way, that watch descriptions can be inaccurate, and that slight imperfections are often not stated. Remember the case metal factor: gold, platinum, silver and steel cases will contribute to different valuations of otherwise similar watch movements. There is also the jewelry factor: an uninteresting watch might have an expensive gold bracelet or its dial might be heavily pavé; there is always a customer for this type of watch but he or she will probably not be a serious collector. Wristwatches described as water-resistant may well not be after multiple pre-sale inspections of movements and so must be re-sealed after purchase.

When the auction commences, take the usual precautions. Often the auctioneer will be bidding himself on behalf of someone, and bidding is often undertaken through saleroom staff by telephone. Unsold lots are not necessarily unwanted or uncollectable; they might have been subject to sellers' reserves. Remember that, in many countries, tax is payable on some prices fetched, or hammer prices as they are called. Note as well that a buyer's premium (which may perhaps be 10 per cent or 12 per cent) is usually added to a hammer price. Finally, as the auction proceeds, always bear in mind that a high proportion of the people around you in the saleroom will be 'in the trade'. The dealers' world is a very small one, and therefore real bargains are very hard to spot and even more difficult to buy. Any watch bought at an auction should be cleaned before wearing.

RIGHT:
Wristwatches sold at auction by Sotheby's, London, on July 24, 1986.

RIGHT:
Wristwatches sold at auction by Christie's, London, on March 3, 1989.

A broad selection of wristwatches which appeared in Sotheby's London saleroom on July 19, 1988. The Ulysse Nardin moonphase (below center) stands out both in quality and style.

Another fine array of collectables from an earlier Sotheby's, London, sale, on February 28, 1985; their buyers today will be most happy at the prices they paid then.

ROLEX
DATEJUST
SUPERLATIVE CHRONOMETER
OFFICIALLY CERTIFIED
CERTIFIED
PRECISION
SWISS MADE
CARTIER
AUTOMATIC
Constellation

LOOKING AFTER WRISTWATCHES

The major rule with an expensive watch is never ever to open the case; the oil will be affected, dust will enter, you will almost inevitably disturb the movement, and it will no longer be water-resistant if it was.

The best place to keep a watch is obviously on your wrist, but beware of sharp-edged cuff-links and bracelets. Do not keep a watch with a broken strap or bracelet in a pocket or a handbag, just so that you can continue using it; scratches come quickly and easily, and severe ones can greatly affect the re-sale value. Wind up your mechanical slowly and carefully, and when your quartz movement battery slows, do not attempt to replace it yourself. Take it to a specialist jeweler or shop and ask them to check the strap or bracelet at the same time. At home, beware of children playing with your watches, if you are storing a spare one, or an evening dress one keep it in a safe or securely hidden.

It is important to read and then keep guarantees from the suppliers. When the time comes to have your watch serviced it is a good idea to take it back to the original retailer; otherwise use good jewelers or local clockmakers with qualifications. Wristwatches prefer, like cars, to be kept in good working order. It is also advisable to test bracelets and straps thoroughly by stretching and inspecting them every so often.

For insurance purposes it is important obviously to keep evidence or the receipts of the original purchase. If you are a serious investor, it is interesting to track auction results, just as you would check stock market prices every so often. Remember that insurance premiums can be too high as well as too low.

Cleaning must be done by the manufacturer or a specialist – any old watch repairer will not do. The best firms have replacement parts for even the oldest models.

LEFT *Watches sold at auction at Sotheby's, London, on October 16, 1986. (Top left) 18 karat gold Rolex DateJust Oyster Perpetual, with its original Jubilee bracelet, (top middle) 9 karat pink gold Rolex Prince, with prominent scrolling lugs, (middle left) 1935 Cartier silver eight day timepiece, (bottom left) gold art deco 1930s watch; case and bracelet in the Egyptian style, (near left) automatic Omega Constellation, 1968, with original gold bracelet.*

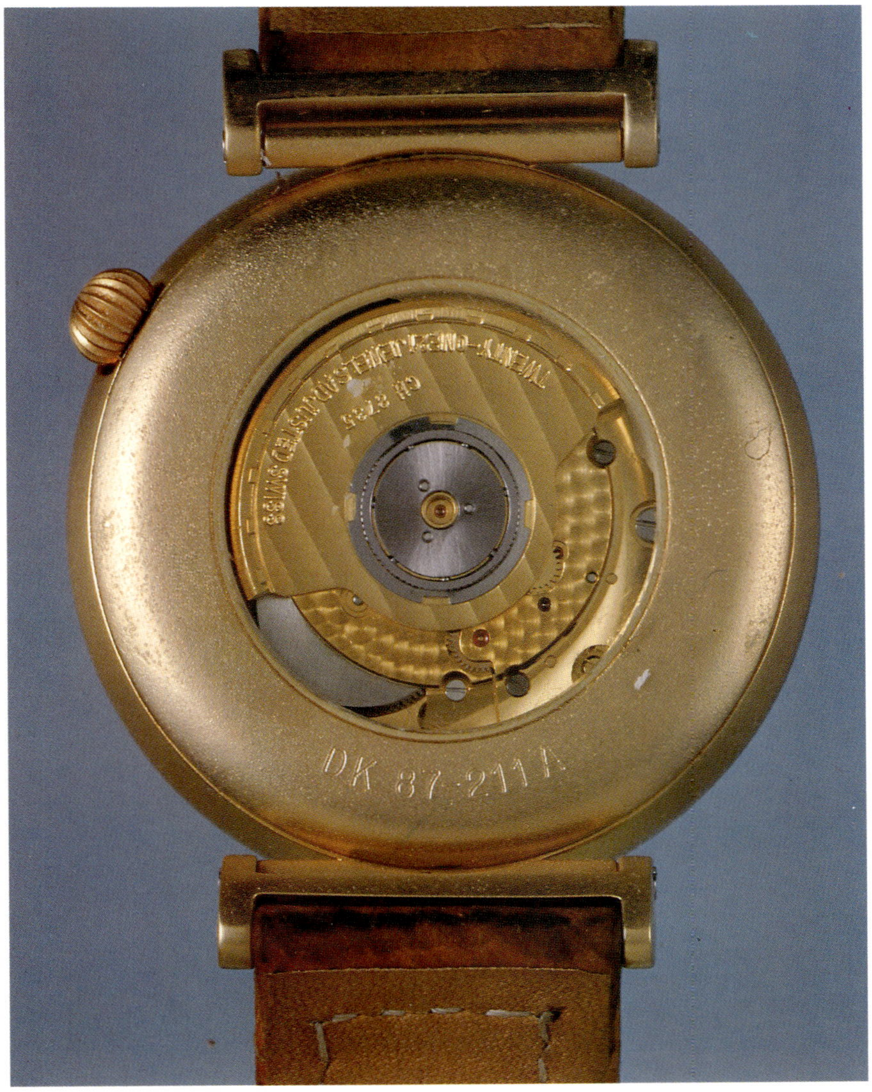

RIGHT *An unusual five-minute repeating automatic wristwatch, in a circular gilt case with hinged lugs, with a white enamel dial, Arabic numerals and gold hands, signed Chronoswiss, Kelek; the glazed sapphire case back is numbered DK87.211A and the 21 jeweled movement within is numbered CH8735 (it repeats on the two gongs, operated by the crown).*

FORMING A COLLECTION

Unless you can afford to collect only classic wristwatches from the great makers, it is probably advisable (as with most good and interesting collections in and outside museums) to choose a cohesive theme for your collection. The one you select will be narrow and finite, you will become very knowledgeable and market-wise about it and finally should you ever decide to sell the collection it should command a premium price because it is ready-made – for someone else to show off or upon which to build further. If indeed you do feel able to begin investing in classic watches, then auction salerooms must be your first ports of call. At first do not bid or buy unless you are employing a specialist dealer to advise and bid for you. Observe and learn.

Only the rarest, most complicated and best makes are likely to show any profit in the short term. When buying 'new' classics remember that the investment potential is mid to long term and relies on the continuing revival of the mechanical and Swiss watch.

The greatest watchmakers today are profiled separately in this book. However there are numerous famous 'names' from the past to look out for, whose watches are increasing in value all the time . . . 'names' such as: Aubry Frères (founded in 1917), Beuchat (1905), Certina (1888), Cyma (1862), Eberhard (1887), Ernest Borel, Eterna (1856), Favre Leuba (1737), Fortis (1912), Le Phare (1888), Meylan (1880), Oris (1904), Paul Ditisheim, Rado (1917), Revue Thommen (1853), Roamer (1888), Rotherham, Tiffany (1837), Universal (1894), Zenith (1865) and Zodiac. This list excludes American makers, who are summarized separately below.

THEME COLLECTING

Here are some ambitious themes for individual collections of wristwatches.

Wristwatches sold at auction by Sotheby's, London, on October 8, 1985.

Wristwatches sold at auction by Sotheby's, Geneva, on November 13, 1984.

CAR WATCHES

These are good fun if you are a car buff, and here are some examples to look out for. A few years ago Aston Martin Lagonda produced very limited editions of watches for each of the marques – some were sold, and the others were given to favored customers. In 1930, Mido produced a Bugatti watch, and later a Buick. Cartier's Ferrari watches are more expensive, but easier to obtain. Bueche Girod long ago used to produce Lancia and Mercedes watches. In 1932, Blancpain made a highly desirable Rolls-Royce wristwatch, in three different models, and Joan Crawford helped promote it. Corum have also made a Rolls-Royce watch: the dial is radiator-shaped; unusually it also came in a ladies' version.

DESIGNER LABEL WATCHES

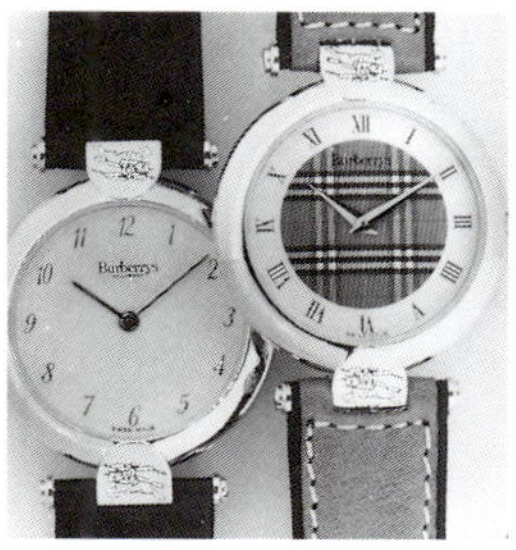

ABOVE *Two Burberrys' own-brand watches.*

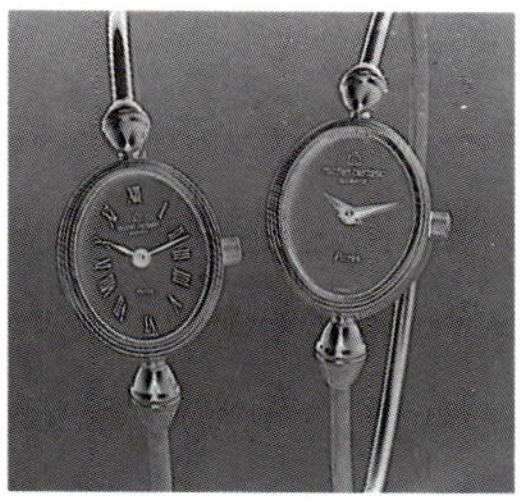

ABOVE *Michel Herbelin invented the bangle bracelet wristwatch as depicted here.*

Money and taste can lead to an amusing collection of wristwatches, featuring the leading names in fashion. Burberrys have watches featuring its distinctive check pattern on either the dial or the strap. Carven and Chanel have ranges. Christian Dior have a beautiful mystery watch called Black Moon; its black dial is covered by a smoked sapphire glass, has no numbers, just gold baton hands and the name, and the strap is matt black sharkskin. Also distinctive for daywear is their Success Rectangular Collection, with its plain white dial, black unusually grouped Roman numerals and baton hands: it is aimed at the Cartier Tank market, but is not expensive. The new Gucci Bombe watches look good, particularly with the mother of pearl dial and a very curved sapphire crystal. Somewhat surprisingly they also offer a chronograph with three subsidiary dials; Gucci have been involved with the Severin Group since 1972. The men's rectangular gold-plated 4200 is also appealing. Guy Laroche has a straightforward collection. Hermès (founded 1837) has had its own watchmaking company in Switzerland since 1978, and some of their models, such as Arcean and Kelly (1978), the Clipper (1981) and the Sellier (1987), are very attractive. Their new (1988) Captain Nemo watch has a unique dial, with its 'swirling' baton numerals. Karl Lagerfeld boasts a wide variety of watches, some surprisingly modern and jokey. Lanvin has some very pretty delicate dials on their watches, which are made for them by Michel Herbelin (this is the company that invented the bangle bracelet). Léonard has their Masterpiece Collection, a coolly elegant line featuring bracelet designs which run through the dials. Pierre Balmain, licensed to SMH, has the dials of the Elegantic line to proclaim. Pierre Cardin has a big inexpensive range, in which the Sportive appeals. Rochas features its name in huge confident lettering on some of its dials. Yves Saint Laurent designs all his watches in Paris, and the ever-changing range is of course brilliantly marketed.

LEFT *Aston Martin Lagonda recently commissioned a limited number of wristwatches each for their two classic car marques; each was serially numbered.*

JEWELERS

Wristwatches and their makers have a long and traditionally very close relationship with jewelers. It is beyond the scope of this book to do more than demonstrate it in the profiles in 'The A to Z of Important Makers' (Cartier, Chopard, Gérald Genta, etc.). Most leading jewelers have 'own brand' watches, but only very wealthy collectors can consider a general collection. An exceptional jeweler/wristwatch retailer is Van Cleef & Arpels; their world timer and skeleton are most notable.

LIMITED EDITION WATCHES

Such watches, in small numbers, are very collectable if they can be found and are affordable. The most majestic pocket watch in a limited edition is Audemars Piguet's La Grande Complication – only 100 have been made since 1915. Cartier, Dunhill, and Mappin & Webb have limited editions, Baume & Mercier's Riviera Complication is one, and so is Movado's Andy Warhol Times/5 (250 only, of which only 200 for sale). Rolex's steel gentlemen's 1940s (ref. 3481), 11 only, and their 1956 Chronographe Antimagnétique (ref. 3055), 100 only, are further examples. In fact most, if not all, the most important makers have produced limited editions, and it is a wonderfully finite and satisfying area in which to explore and invest.

MYSTERY WATCHES

A mystery watch has no numerals on its dial and only two hands; it is a pleasing category in which to collect because such watches are available at all price levels. A 1950 Jaeger-LeCoultre gold mystery watch fetched 4,180 Swiss francs in November 1988 (Sotheby's, Geneva); this company has made several through the years. Movado's Museum watch is famous (and unobtainable, it is said). Cartier and Piaget are among other makers to have made them, and they feature continually in the auction salerooms. Gérald Genta's Secret Time is a very fine contemporary example of the state of the art. Early pocket watches had only one hand; Revue Thommen has a wristwatch with no hour hand: how many people do not know the hour?

ABOVE *An eminently modern watch in two versions from a traditional jeweler, Tiffany & Co.*

OPPOSITE *A watch for left-handers. The crown is on the left of the case in this quartz Accurist wristwatch with moonphase and three subsidiary dials.*

Accurist
QUARTZ

NOVELTY WATCHES

There will always be a market for the out-of-the-ordinary and the wildly eccentric . . . just so long as the watches tell the time. Salvador Dali's famous painting *The Persistence of Memory* (1931), and his other watch paintings, have inspired a number of watches. The artist himself is on record as saying that it is 'nothing else than the tender, extravagant and solitary paranoiac-critical Camembert of time and space'. All very clear! The Italian goldsmiths and watchmakers Dino and Roberto Falcone have made three such watches, each different. The Cartier Dali watch (1970s) is more impressive; one fetched 40,000 Swiss francs in June 1988.

There is a more contemporary range of watches, some of which are classics in their own right, in spite of their numbers and low prices. These, of course, are the ubiquitous Swatches, and their story is a fascinating one. On April 11 1980, ETA (part of the giant Swiss SMH group) decided to develop an 'economic' watch to combat the Japanese digital watches, which were then causing the closure of many smaller Swiss manufacturers, and causing larger ones to amalgamate. In 1981 the idea of Swatch was conceived, and by 1983 it was on the market in USA and Britain. By September 1988, 50 million pieces had been sold worldwide and 406 designs had been produced. In the unique manufacturing process 51 parts are assembled on the baseplate and then sealed into a water-resistant case, which itself is as important as the dial in terms of design; Swatch straps and guards bring further individuality to the 'wrist package'. And every Swatch has a name; the marketing concept is brilliant. From the collector's point of view it is possible to collect the whole lot, and they sell a smart case to put them in. One favorite is the 1985 Jelly Fish.

The earliest novelty wristwatch was Ingersoll's Mickey Mouse (1933) and two million were sold in three years; it is said that the late Emperor Hirohito wore one for about 50 years, and it is increasing in value all the time. Timex also produced a Mickey Mouse watch. Apollo (well-known for their watchstraps, made of all kinds of material, including chicken leg skin) sell a contemporary Mickey Mouse in their Disney Collection. The USA has produced many classic novelty wristwatches over the years, showing characters such as Popeye, Winnie the Pooh, Donald Duck, Minnie Mouse, Goofy, Bambi, Annie Oakley, Li'l Abner, Roy Rogers and Trigger, Davy Crockett, Zorro, Hopalong Cassidy, Lone Ranger, Woody Woodpecker, Batman, Superman, Wonder Woman, Cinderella, Pinocchio, Alice in Wonderland, Space Cadet, and so on. The leading manufacturers included Ingersoll, Timex, US Time, New Haven, and particularly E. Ingraham Co. (USA).

LEFT AND RIGHT *Flik-Flak watches.*

RIGHT: *Seiko's remarkable Speech-Synthesizer. This digital quartz has a recording and play-back capability in four or eight seconds, chronograph, alarm, automatic month, day and date calendar and an illumination light.*

RIGHT: *A range of classic pop Swatch wristwatches in an advertising mode.*

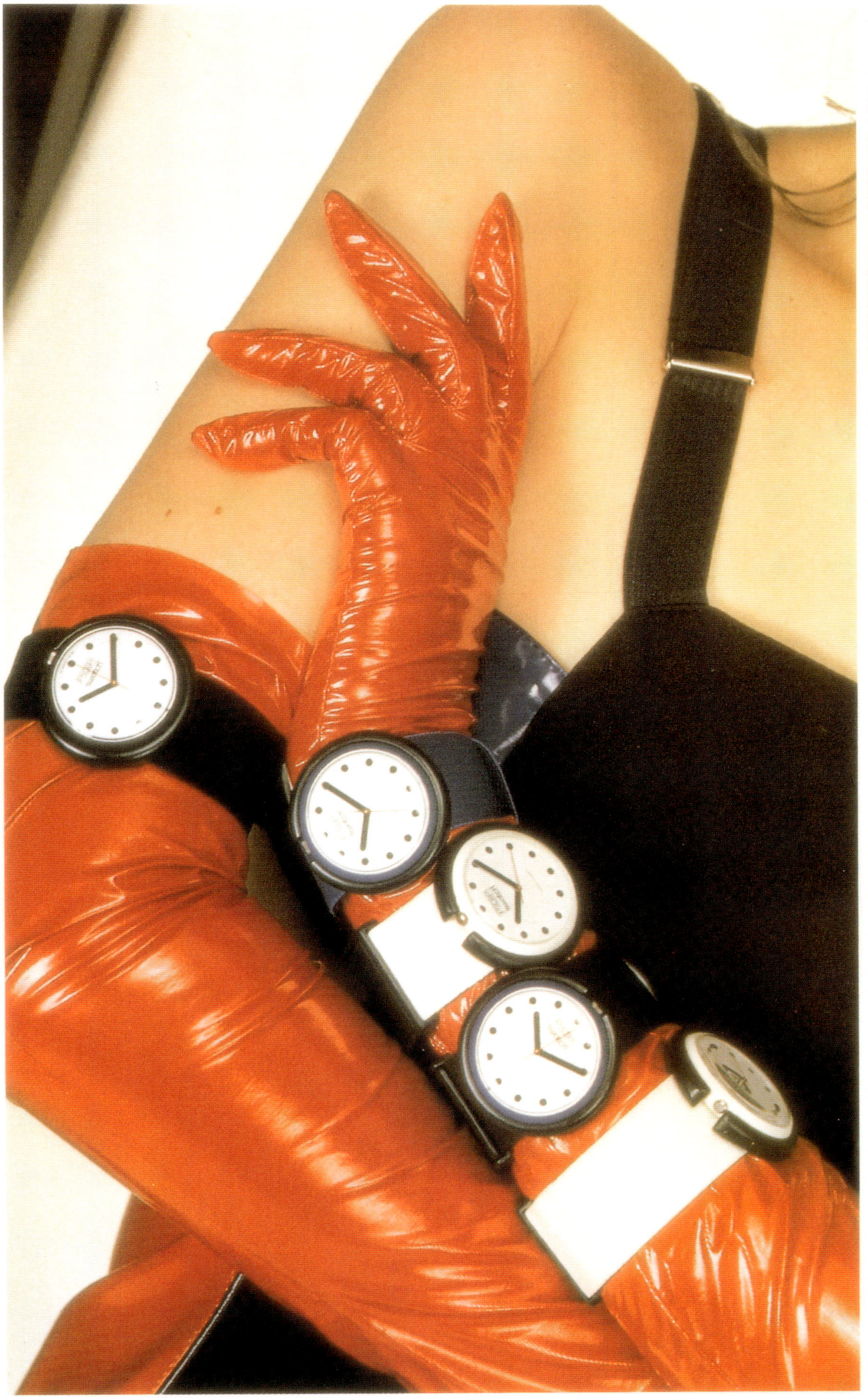

TAG
HEUER

There are a few 'left-handed' watches around, with crowns on the left. Accurist have one currently available; in the 1920s there was such an English watch called Dublin. Watches with hologram dials are about. Two adult novelties, which will be future collectables, are Fiorucci-Time Roulette; it has an integrated and functional roulette mechanism, based on a ball-bearing system. A push button above the crown activates the numbered disk within the bezel. Also for adults, especially if you win at roulette, Swiss Tonic (a Challenger product) have recently brought on to the market a wristwatch, from the strap of which the case can be unhinged to reveal in the recess beneath either a condom or pills. Watches with wooden cases have been a specialty of Lacher (based in Pforzheim) for 60 years; Timber in France have an elegant more expensive range which is worth watching for.

ABOVE *Lacher's Wood Watch, one of a wide range.*

SPORTS AND HOBBIES WATCHES

OPPOSITE *TAG-Heuer Formula 1 chronographs offer advanced materials, fashionable color combinations and also an exciting range of functions which mark them out as contemporary classics among sports watches.*

The great outdoors produces its own demanding sets of requirements where time-keeping is concerned, and the wristwatch market in recent years has addressed itself in increasingly sophisticated ways to its strict demands. Several of their products will become classics. The rough and tumble of adventure pursuits such as car rallying are well served by Breitling (see the profile in 'The A to Z of Important Makers'), Revue Thommen and Tag-Heuer. Deep sea divers can turn to Citizen's remarkable Aqualand range, complete with depth sensors and alarms, wind velocity dials, etc. Tag-Heuer's Series 1000/2000/3000 are designed for professionals. Up above, sailors and wind-surfers can choose from Tag-Heuer's colorful steel-cased range of chronographs with their colored bezels. Krieger, in Florida, have a remarkable Tidal Chronometer; this wristwatch shows the present state of the tides. The automatic clockwise rotation and position of the red ball on the tide display, driven by microscopic gears, indicates the tidal position at any particular port. The graduations on the tide display instantly give you the present state of the tides at that time and place, and it is possible to count up the hours until the next high or low tide. The red ball also duplicates the shape of the moon; two rotating disks show the moon moving from full to a sliver, so that you can forecast Spring and Neap tides. Citizen have wonderfully complicated analog, with digits, watches for this market. Citizen also offer wristwatches for mountain climbers (with elevation sensors), for cyclists, skiers and even parachutists. And of course they have the Wingman, a pilots' watch. Collectors already know of the most famous early aviators' watch, Longines' Charles Lindbergh (one fetched $8,415 in October 1988 at Sotheby's, London). The same value at that sale was put on a rare frogmen's Rolex. Collectors should look out for Universal's 18 karat Lady's Bridge watch. Again Gērald Genta is at hand, with investments for the future – his Gefica Safari for hunters, and L'Open for golfers. Revue Thommen also have an admirable gold golf scoring watch.

OVERLEAF LEFT: *Perhaps this modern watch from Citizen will come to be regarded as a late 1980s classic in years to come. It is the Altichron and is the world's first watch with an elevation sensor (extending to the left) and compass, for climbers. It can only be used in the Northern hemisphere, and is available in meter and feet models.*

OVERLEAF RIGHT: *TAG-Heuer's s/el diver's watch which has been given a smoothed-off bezel and highly fashioned lugs, giving it a certain functional elegance. It is water-resistant to 200m/660ft.*

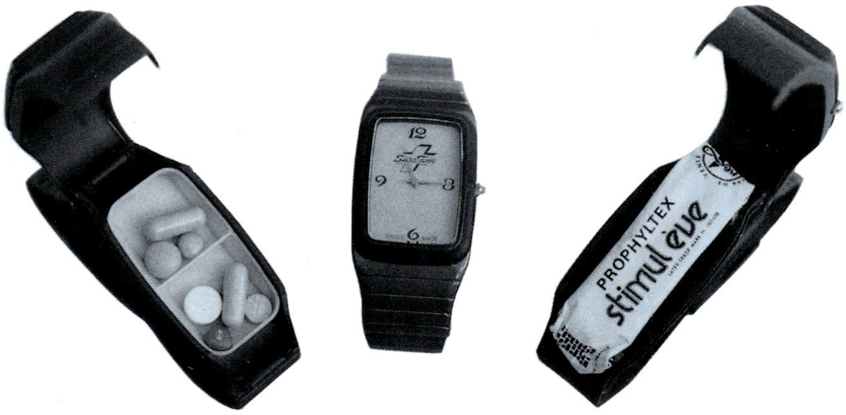

Swiss Tonic's hinged quartz wristwatch can contain such necessities as pills and condoms.

CITIZEN
QUARTZ
WATER RESIST

TACHYMETRE
TAG HEUER
TACHYMETRE
TAG HEUER

OTHER THEMES

There is naturally a myriad of themes to suggest; here are further wristwatch categories: alarms, Art Deco, watches with bangle bracelets, masonic, mechanical digital, military, skeletons, and world timers.

ABOVE *A gold and steel Rolex Oyster Perpetual watch with subsidiary seconds. The most collectable dial name at the generally affordable level.*

RIGHT *A typical example of a World War I military wristwatch, in the style of a pierced half-hunter.*

OPPOSITE *The beauty of a Lemania skeleton movement.*

BREGUET 0001

GLOSSARY OF WATCH MAKING TERMS

ARABIC NUMBERS
0, 1, 2, 3 ,4 ,5, 6, 7, 8, 9. Originated in India and introduced by the Arabs to Europe in about the.10th century AD.

ANALOG (ANALOGUE)
Time indication by hands and dial; means 'corresponding'. Originally an electronic term, which was adopted into watch-making with the spread of the quartz watch.

ANTIMAGNETIC WATCH
Watch whose parts are protected from all but the very strongest magnetism; quartz watches cannot be disturbed by the phenomenon.

APPLIED NUMERALS
Raised metal characters attached to the dial.

AUTOMATIC WATCH
Mechanical watch with a mainspring that is wound by the wearer's movements, via a rotor. Invented by Abraham-Louis Perrelet in the 18th century; Breguet called his self-winders 'perpetuelle'.

AUXILIARY DIAL
Small dial showing seconds only, up to one minute, usually at the six o'clock position.

BACK PLATE
Furthest from the dial.

BACK WINDER
Flat crown set into the back of the case for setting time and winding.

BAGUETTE
Rectangular movement, with a length at least three times its width. Popular shape for art deco watches.

BALANCE
Running regulator of mechanical watch; it oscillates about its axis of rotation, the hair-spring making it swing to and fro ('tick-tock') in equal time parts. Balances of modern wristwatches beat up to ten beats per second.

BAND
Center ring of wristwatch case into which the movement fits; front and back bezels are hinged to it.

BARREL
Circular box housing mainspring; teeth attached at edge drive gears; going barrel has great wheel mounted upon it.

BATON NUMERALS
Undecorated non-numerical markers of hours, minutes and seconds.

BEZEL
Metal surround frame in which watch glass (crystal) is fitted.

BI-METALLIC
Two different metals (eg brass and steel) fused or riveted together to make a single strip, with dissimilar coefficients of expansion, for balances; this prevents variations in temperature affecting the watch mechanism.

BREGUET HAND
Popular design by Breguet; the slightly tapered needle of the hand ends in a pointed head mounted on a circle, which is pierced with a hole. Sometimes called a moon hand.

BRIDGE
Strips of metal that hold pivots or jewels in place in the movement.

BUTTON
Better known as a crown, or winder; sometimes refers to chronograph.

CABOCHON CROWN (WINDER)
Crown or winder set with a jewel.

CALIBER (CALIBRE)
Once used only to denote the size in diameter of a watch movement; now often only indicates type (eg men's, ladies', automatic). Generally given with manufacturer's name. From Latin *qua libra* (of what weight?), or from Arabic *kalib* (mould, ie, circumference, measurement, scale).

CASE
The housing for movement, dial and glass.

CENTER SECONDS
Second hand rotating at center of dial, concentric with hour and minute hands, indicating seconds at edge of dial. Also known as sweep seconds.

CHAPTER RING
Circle of hour divisions indicators.

CHRONOGRAPH
Watch which also has an independent stop-watch for short interval timing. Common types are one-button (using crown, or separate button above it); two-button (the most common, the top button stopping and starting the time-measuring function and the bottom one resetting it); 12 hour with moon phase; split-second.

CHRONOMETER
Ordinary watch which has passed extremely severe precision and reliability tests in an official (generally Swiss) observatory (eg Neuchâtel).

COMPLICATED WATCH
Watch with functions not related directly to the time of day (eg calendars, chronographs, moonphases, perpetual, repeaters, etc).

CROWN
Knob, generally knurled and positioned outside the case at three o'clock, for winding, correcting and setting.

CRYSTAL
Glass dial cover (in fact made of glass, plastic, synthetic sapphire or quartz crystal), fitted into bezel. Plastic scratches; glass (common in pre-1940s watches) shatters easily; sapphire glass is virtually scratch-proof.

DATE APERTURE
Window showing date in mechanical digital watches.

DEPLOYMENT BUCKLE
Two strips of hinged metal (curved to the wrist shape) on the watchband; upon closing, one folds over the other to cover it. Probably invented by Cartier.

DEVIATION
Time discrepancy from the actual.

DIAL
Face of watch, showing hours, minutes and seconds. Other small dials are called subsidiary dials.

DIAMETER
Dimension used in measuring a watch size.

DIGITAL NUMBERS
Indication of time displayed by number only.

DIVER'S WATCH
Water-resistant.

DOCTOR'S WATCH
Also known as a duo-plan or duo-dial. An auxiliary seconds dial is separate from the hour and minute dial; useful for quick reference when taking a pulse count.

DOME
Second cover inside the back of a watch.

EBAUCHE
French word remaining untranslated and in common use. It describes an uncompleted watch movement delivered to a manufacturer ready for the addition of that manufacturer's own escapement, timing system and mainspring and signature.

ELECTRIC WATCH
Watch incorporating electric contacts, coils, condensers or resisters, in which closed circuits balance impulse. Hamilton and Elgin (USA) electric watches first appeared in 1952; Lip in France also make them.

ELECTRONIC WATCH
Quartz watch, with semi-conductive elements like transistors.

ESCAPE WHEEL
The last wheel in the going train; alternately locked by and gives impulse to the balance.

ESCAPEMENT
The parts of the movement that convert the rotary motion of the gear train into exact 'to-and-fro' motion.

FORM WATCH
Watch in any very unusual shape.

FORMED MOVEMENT
Watch movement that is not round (eg baguette, oval, rectangular).

FRAME
Pillars and plates of watch movement.

FREQUENCY
Number, generally expressed as hertz, of cycles, oscillations, periods or vibrations per second.

GEARS
Toothed wheels (20–100 teeth, of brass) and pinions (six to 12 teeth, of hardened steel), running together.

GOLD
Yellow, pink or white, used for cases and bracelets.

GREAT WHEEL
First and largest in train.

HACK FEATURES (BALANCE STOPPING)
Second hand which is stopped to synchronize time, when crown is pulled out.

HALF-HUNTER
Glass half covered by hinged case extension.

HERTZ
The convention for expressing the number of cycles, oscillations, periods or vibrations per second. Abbreviated to Hz, KHz (1,000 Hz) or MHz (1,000,000 Hz). Named after the German physicist Heinrich Rudolf Hertz (1857–94).

HUNTER
Glass wholly covered by hinged case extension (often spring-loaded). Also known as savonette watches.

INTEGRAL BRACELET
Designed as a natural extension of watchcase.

JEWELS
Used as bearings at points of greatest friction in movements; commonly 15 to 18 are used (the quantity is not indicative of either quality, or value of watch). Formerly, natural rubies and sapphires were used; today most such jewels are synthetic.

JUMP HOUR
Hour hand which moves forward only once each hour; alternatively hour appears only once each hour in a pierced window.

KARAT (CARAT)
The official scale by which the purity of the gold is determined. Pure gold is 24 karat; 18 karat is alloy in which 18 parts in 24 are gold; 14 karat contains 14 parts of gold and so on. Also used as the unit of weight for precious stones.

LCD
Liquid crystal display, in almost all quartz watches.

LEPINE CALIBER
Caliber in which movement has only one plate, to which each wheel is supported by a separate cock. Invented by Jean Antoine Lepine (1720–1814); perfected by Breguet for ultra-thin watches.

LEVER
Brass or steel escapement part of mechanical watch, impulsing at each vibration, shaped like a ship's anchor. Invented by Thomas Mudge (1715–94) in 1754 and now universally used.

LIGNE
One inch is one ligne, or one-twelfth of a French foot (approx 2.256cm). Lignes are commonly used for giving wristwatch movement sizes (diameter, length or breadth), the commonest being between 5,5 and 13.

LUG
Part or parts of watch case to which band, bracelet or strap may be attached.

MAINSPRING
Principal spring in watch; a flat spring is coiled in a barrel.

MEAN TIME
Average length of all solar days in year; the usual time shown by watches.

MINUTE REPEATER
Repeating watch that sounds hours, quarters and minutes.

MONTH APERTURE
Pierced window in a mechanical digital watch displaying month, often abbreviated.

MOON PHASE WATCH
Watch displaying phase of moon through 29.5 days (correction for extra 44 minutes per month often incorporated).

MOTION WORK
Gear train for moving hour hand.

MOVEMENT
Complete mechanism of watch; from 120 to over 600 parts may be incorporated in it.

OSCILLATION
'To-and-fro' swing between two extreme positions (see also BALANCE).

OYSTER CASE
Rolex watch with water-resistant case.

PAVE
Literally 'paved with', as in dial with precious stones.

PALLET
Small jewel for locking escape wheels and receiving impulses.

PALLET FORK
Jewel-tipped lever in escapement, in conjunction with balance and escape wheel.

PERPETUAL
Self-winding automatic watch (see also AUTOMATIC WATCH).

PERPETUAL CALENDAR
Calendar mechanism with display which automatically corrects for long and short months and leap years. Formula adjustments for vagaries of the Gregorian calendar continue only until February 28, 2100; that is not a leap year, so manual changes will have to be made to all but the most complicated watches; likewise 2200, 2300, 2500, 2600 and 2700 will not be leap years.

PLATE
Parallel flat plates in which wheels of train are pivoted.

PLATINUM
Precious silver-white metal, which is heavier than gold. Used for cases and bracelets.

QUARTER-REPEATER
Repeating mechanism which sounds hours and quarter hours.

QUARTZ
Rock crystal (silicon dioxide) that can be made to oscillate by electronic switching, maintaining its very constant frequency, in accordance with its cut. Synthetic quartz crystals are used today.

ROLLED GOLD
An extremely thin sheet of hot gold, pressed on to another metal; gold on watch cases is usually double thickness.

ROMAN NUMERALS
Besides Arabic, the most common numerals used on watch dials; note IIII instead of IV.

ROSKOPF WATCH
Simplified mechanical watch invented in 1867 at La Chaux-de-Fonds by G.F. Roskopf (1813–89); usually without jewels and with an unusual gear train.

ROTOR
In automatic watches, the rotor winds the mainspring; in quartz watches, it is a permanently rotating magnet in the step-switch motor.

RUBY
The 'ruby' referred to in watch-making today is, in fact, corundum, a synthetic stone. It is used to reduce wear on certain pivots.

SAPPHIRE
Glasses, sold as scratchproof, are made of synthetic sapphire.

SAVONETTE
European term for hunter watch.

SHOCK-RESISTANT WATCH
A watch is held to be shock-resistant if, when dropped on to a hardwood surface from a height of 3ft (1m) it does not stop, or if its daily rate does not change by more than 60 seconds.

SIDEREAL TIME
Standard of time used by astronomers; the sidereal day is three minutes and 55.5 seconds shorter than a mean solar day.

SIGNED MOVEMENT
The signature on a movement of its maker, which is likely not to be the same as that on the dial.

SKELETON WATCH
The dial of a skeleton watch has a separate chapter ring with the interior cut away, leaving only numerals and exposing the wheels and the interior mechanisms of the movement. The back plate is also cut away and fitted with glass.

SOLAR TIME
As shown by a sundial.

SPLIT SECOND CHRONOGRAPH
Chronograph with sweep second hand, independent of chronograph hand.

STEM
Shaft connection between winding mechanism and crown on outside of case.

STOP WORK
Device controlling number of winding turns, thus preventing overwinding.

SUBSIDIARY DIALS
Smaller auxiliary dials, which show elapsed minutes and running seconds.

SWEEP SECONDS (CENTER SECONDS)
Second hand mounted at dial center and extending to chapter ring.

'SWISS MADE'
A Swiss Federal government ordinance dated December 23, 1971 decrees that this expression can only be featured on a watch and used in connection with its marketing if (a) at least 50 per cent of the components, by value, excluding costs of assembly, are of Swiss manufacture, (b) it was assembled in Switzerland, (c) it was started up and regulated by its manufacturer in Switzerland, and (d) it is continuously subject to the legal obligation of technical inspection in Switzerland.

TACHOMETER
Speedometer or revolution recorder on bezel.

TANK CASE
Today, the common name for a rectangular case; originally, exclusive name of Cartier wristwatch.

TIMEPIECE
Domestic clock that does not strike or chime.

TONNEAU
Case shape with wide center and flat tapered ends.

TOURBILLON
Invention by Breguet for nullifying vertical position errors by means of a revolving platform which goes through all such positions, so that they neutralize each other.

TRAIN
Wheels and pinions of a watch, carrying power from the great (first) wheel to the escapement. Variations in their numbers of teeth affect the running time and number of oscillations.

TRITIUM
Luminous paint for dials, hands and numerals.

TUNING FORK
A transistor continually switching between two small magnets to regulate smooth running, oscillating 360 times a second. The high frequency gives great precision in time-keeping. Bulova Accutron made the use of the device famous, but then quartz watches usurped its popularity.

WATER-RESISTANT
Expression for 'waterproof', which is illegal in the USA. Water-resistant watches, sold as such, must be able to withstand water pressure at a depth of 1m (3.28ft) for 30 minutes and thereafter for 90 seconds at 20m (65.6ft). Divers' watches in fact have much greater resistance.

WHEEL TRAIN
In a normal wristwatch, first (great) wheel, which supplies power; second (center) wheel, turning once per hour and carrying minute hand; third wheel, which transmits power to the fourth wheel, which sometimes carries second hand; escape wheel.

WORLD TIME WATCH
A watch that can be made to depict current time in any chosen city or zone, according to the model.

INDEX

ACKNOWLEDGMENTS

I have gratefully received all kinds of help during the writing of this book. My requests for information from manufacturers were met with prompt supplies of watch details and historic backgrounds; I sincerely thank them all.

I was particularly glad to have the active support and encouragement of the following: Geoffrey Ashworth, Grahame Brooks, George Daniels, Terry Davidson, Roger Day, Mahmad Dilloo, Edward Dreyfuss, John Goodall, Bob Hawkesworth, John Keeping, J.D. Metcalfe, Robert Peppiatt, Henry Polissack, Salwa Rusconi, the Scheufele family, Rolf Schnyder, Rita Shenton, George Somlo, Joan Stemp, Sally Stevens, Stewart Unger, Trevor Wayman and Philip Whyte.

Marcus Margulies was greatly helpful with his introductions, and both David Briercliffe and Frank Edwards gave me much time and valued advice. Karen Mount handled the substantial volume of correspondence and manuscript typing with typical cheerfulness and efficiency.

Leading auction houses were tremendously helpful. Sincere thanks to Richard Price (Bonhams), Simon Bull, (Christie's), Roger Lister (Christie's, South Kensington), Ben Wright (Christie's, King Street), Oswaldo Patrizzi (Habsburg, Feldman, Geneva), Christopher Greenwood (Phillips, London) and Tina Millar (Sotheby's London).

Finally, huge thanks to Elisabeth Ingles, who has given me immeasurable support in so many ways.

MICHAEL BALFOUR

KEY t=top; b=bottom; l=left; r=right; c=center

The author and publishers have made every effort to identify the copyright owners of the pictures used in this publication; they apologize for any omissions and would like to thank the following:

Accurist: 175; Apollo: 18; Aston Martin: 173b; Audemars Piguet: 28, 31; Bertolucci: 34, 35; Breitling: 44, 45, 46, 47; Bridgeman Art Library: 157; Burberrys: 173l; Cartier: 48b, 51, 52, 53; Chopard: 54, 55, 56, 57; Christie's Colour Library: 10, 16, 33, 42, 49, 61t, 61c, 93l, 100, 114, 121b, 127t, 128, 129, 137, 138, 139, 142; Citizen: 180; Corum: 59, 60, 61; Daily Telegraph: 11; © Demart Pro Art B V/Dacs 1989: 20–1; Dunhill: 62, 63; Ebel: 68; Flik-Flak: 176t; Georg Jensen: 69, 70, 71, 72; Girard Perregaux: 75, 78, 79; Habsburg, Feldman: 10; Ingersoll: 82; International Watch Company: 83, 84–5, 90; James Bond © 1973 Danjaq S.A. All rights reserved: 141; Jaeger-LeCoultre: 92, 93r 94b, 95, 96; Jean d'Erc: 17; Lacher: 179t; Lip: 23; Longines: 97, 102, 103; Mary Evans Picture Library: 12, 19; Michel Herbelin: 173r; Mido: 104, 105, 106, 107; Movado: 108-9, 110, 111; Omega: 113, 115, 116, 117, 118, 162t; Patek Philippe: 119; Raymond Weil: 130, 131; Rolex: 133, 136; Rotary: 143; Seiko: 176b; Sotheby's Auctioneers, London: 30, 48t, 50, 67, 94t, 109b, 120, 121t, 126, 132, 134-5, 163, 172; Swatch: 22, 177; Swiss Tonic: 179b; Tag Heuer: 178, 181; Tiffany & Co: 162b, 174; Time Museum: 74, 159; Tissot: 144, 145, 146; Ulysse Nardin: 147, 148, 150; Vacheron Constantin: 151